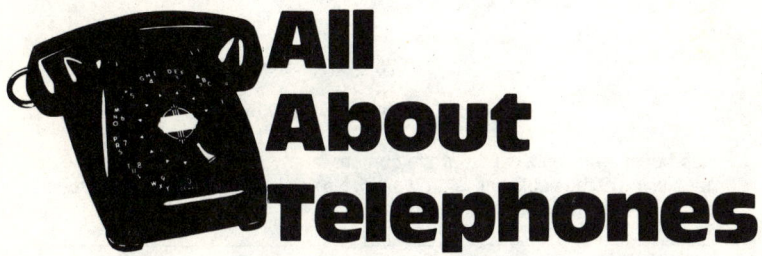

By Van Waterford

BLUE RIDGE SUMMIT, PA. 17214

FIRST EDITION

FIRST PRINTING—DECEMBER 1978

Copyright © 1978 by TAB BOOKS

Printed in the United States of America

Reproduction or publication of the content in any manner, without express permission of the publisher, is prohibited. No liability is assumed with respect to the use of the information herein.

Library of Congress Cataloging in Publication Data
Waterford, Van.
   All about telephones.

   Includes index.
   1. Telephone. I. Title.
TK6165.W37   621.385   78-10574
ISBN 0-8306-9864-7
ISBN 0-8306-1097-9 pbk.

All About Telephones

Other TAB books by the author

No. 1052 *Radar Detector Handy Manual*

Contents

Introduction 7

1 Development of the Telephone 11
Early Discoveries—Bell's Experiments—The Bell System—GTE Automatic Electric Company—ITT: International Telephone Telegraph—Today's Telephone

2 How Your Telephone Works 27
The Loop—The Telephone Set—The Switching Network—Hooking Up Your Own Phone

3 Analog vs. Digital Transmissions 44
Lightwave Communications—Satellite Communications—Computers and the Telephone

4 FCC Rules and Regulations 61
Manufacturer and Supplier Registration—Owner Requirements—Telephone Company Requirements

5 The New Telephones and Accessories 68
The Picturephone—The Speaker Phone—The Electronic Telephone — Cordless Telephones — Miscellaneous Accessories—Telephone Answering Machines

6 Decorator Telephones 110
GTE's Decorator Phones and Accessories—AT&T's Decorator Phones and Accessories—ITT's Decorator Phones—Other Decorator Phones

7 Telephone Security Devices 128
Voice Scrambling Devices—Wire Tap Debuggers—Voice Stress Analyzer System—Security Alarm Systems—Automatic Telephone Dialer

8 **Facsimile Communications** 142
Stringent Requirements—Early Facsimile Methods—The Photoelectric Cell—Electronic Mail—Scrambling Facsimile Communications

9 **Mobile Radiotelephones** 148
How Mobile Telephones Work—Station to Mobile Unit Calling—Mobile Unit to Base Station Calling—Installation Technique—AttachePhone—The Skyphone—Government Regulations—The Radio Common Carrier—Computer Technology and the Mobile Phone

10 **The Future** 167

Appendix I Glossary 173

Appendix II List of Suppliers 182

Index 189

Introduction

We Americans talk more over the telephone than any other nation. Wherever we congegrate in an office building, a stock exchange, a club, a university, a hospital or hotel, in every factory or newspaper building, telephones are installed to help us communicate faster and farther.

Next to the Bible, phone directories are read more than any other publication—more than 100 million people use the Yellow Pages over 4 billion times a year. Surely, impressive statistics.

Our American ingenuity and technological know-how have helped us to become very "communications conscious." There are a lot of things you can do with your telephone today. The advanced technology of communications aided by transistorized electronics and computers has made the telephone a versatile instrument which works quickly and is easy to use.

The Bell System some years ago undertook the biggest project in its history, creating Electronic Switching Systems, ESS for short. The new system does all switching electronically and faster, using only a fraction of the power and space needed for the old electromechanical gear. This electronic system can also be instructed to let you transfer calls, bring a third party into the conversation, be given a signal while you're talking that someone else is trying to call, or even let you dial a frequently used number with only two digits.

Besides striving for faster, more convenient telephone service, telephone companies have assisted the handicapped by designing automatic dialers and hands-free telephones with speakers. Those who are hard of hearing can get phones with a volume control dial to amplify the ring and the conversation.

Many services are now available through the telephone system. TV programs, drawings, photographs, computer data and diagrams can all be sent over the telephone network now. All of these services come to you in a wide range of styles, shapes and colors.

Chapter One traces the history of the telephone, from Alexander Graham Bell's invention, to Bell Lab's many achievements. Also described are the histories of such giants as GTE Automatic Electric (the largest independent phone company) and ITT.

Chapter Two describes the workings of your telephone: the transmitter/receiver, the hookswitch, the bell ringer. This chapter also deals briefly with the workings of switching networks and the varieties of telephone exchanges.

Chapter Three describes the differences between analog and digital transmissions, the new laser lightwave guides used by the telephone company, the relationships between the phone and computers, and the phone and satellites.

Chapter Four informs you about the new FCC regulations and how they apply to the phone supplier, to you the user, and to the telephone company. This chapter clarifies who is responsible for what in the confusing world of telecommunications.

Chapter Five describes, often in detail, all the new telephones that are now available, and their accessories: the Picturephone, the hands-free speaker phone, a computerized telephone, cost-saving devices that let you immediately determine your telephoning costs, cordless telephones, automatic dialers, and telephone answering machines.

Chapter Six describes the variety of decorator telephones now on the market. A large selection of photographs illustrates various designs and shapes available.

Chapter Seven informs you about the devices that scramble your phone conversations so that no one who might be bugging your line will understand a word of what's being said. Of course, your party on the other end, with the proper equipment, will hear every word you say. You can also debug your telephone line in case you suspect telephone wiretapping. This chapter also shows you how to install burglar and fire alarms, and connect these to an automatic phone dialer. When you're not home, this dialer can automatically inform you, or other designated parties, about the situation at your house.

Would you like to send a recipe fast, without having to dictate it to the party on the other end who has to write it down? Use a facsimile device. Chapter Eight deals with these units that allow you to send written materials, or the picture of your latest grandchild, over the phone lines.

In case you want to install a mobile telephone in your vehicle, Chapter Nine explains in detail how this type of telephone works, how to install it, and how to operate the controls. Also described are a portable attache case telephone, one that you can take along wherever you go (a forerunner of the wristwatch phone?), and a Skyphone, in case you want to make a phone call when flying in a private plane.

The final chapter, Chapter 10, deals with the future of the telephone...the wristwatch telephone, satellite communications, and computerized switching.

I would like to express a special word of thanks to Mr. Anthony W. Hilvers of GTE Automatic Electric for all the support given to me in order to make this book as complete as possible. Words of special gratitude are also in order to Messrs. Bob Katzeff, of AT&T, and Jes E. Schlaikjer of ITT.

This book would not have been possible, of course, without the material and information so many suppliers and manufacturers have graciously submitted—their support is greatly appreciated.

I hope that by reading this book and about the devices you may purchase, you will save money and time with your increased communications possibilities.

If you have any questions about your phone service, or any of the units described, call or write to the business office of your telephone company.

Van Waterford

Development of the Telephone

In 1874, Alexander Graham Bell insisted, "If I can make a deaf-mute talk, I can make iron talk." The stubbornness of this young Scottish speech teacher, untutored in electricity, led him to invent, not one, but two practical telephone designs by 1876.

Although Bell finally succeeded in "making iron talk," several scientists before Bell had made preliminary observations about voice transmission.

The English scientist Robert Hooke (1635-1703) made the first suggestions of how speech might be transmitted over long distances. After some experiments on the transmission of sound over taut wires, he remarked, "Tis not impossible to hear a whisper a furlong's distance, 201 meters, it having already been done; and perhaps the nature of the thing would not make it more impossible though that furlong should be ten times multiplied."

In 1796 the German scientist G. Huth suggested that acoustical telephony might be tried. He had the impossible idea that during clear nights mouth trumpets or speaking tubes could be used to pass a shouted message from tower to tower. Optical telegraph towers were about ten kilometers from each other, so this could never have worked. However, Huth is remembered for one sentence in his book, "To give a different name to telegraphic communications by means of a speaking tube, what could be better than the word derived from the Greek *telephone*."

EARLY DISCOVERIES

The incentive to invent the electrical telephone was by no means as pressing as that of the electrical telegraph. The need for relatively rapid transmission of messages having been satisfied, fewer men worked on the development of the telephone than on the telegraph. There was, for example, the American physician, Charles Grafton Page (1812-1868) of Salem, Massachusetts, who discovered in 1837 that when there are rapid changes in the magnetism of iron, it gives out a musical note; the pitch of the note depends on the frequency in the changes of magnetisation, and he called these sounds "galvanic music."

Many physicists repeated these experiments in their laboratories, but it remained for Philip Reis (1834-1874) of Friedrichsdorf, near Frankfurt-am-Main in Germany, to be the first in 1860 to transmit a musical melody electrically over a distance. He stretched an animal membrane to which a platinum wire was fastened by means of sealing wax over a small cone in the shape of a human ear, and inserted this into the bung hole of a beer barrel. The platinum wire formed part of a battery circuit with a corresponding frequency. At the other end of the circuit, the wires led to a coil which was wound around a knitting needle. These rapid magnetisations and demagnetisations of the knitting needle reproduced the sound. Three years later he claimed that words could also be made out.

Two Americans were working on telephonic transmission in 1875, independently of and unbeknowns to each other. One was Elisha Gray (1835-1901); the other was Alexander Graham Bell (1847-1922). Gray was an inventor and a manufacturer in Chicago. His telephone was not unlike that of Reis, but he attached a small iron rod to his membrane, and the other end of the rod was immersed in a fluid of low electrical conductivity, part of a battery circuit. A sound reached the membrane, vibrated the rod immersed in the fluid, and passed a fluctuating current down the circuit. On the receiving end, the wires of the circuit passed into the coil of an electromagnet enclosing another small rod of soft iron, also attached to a membrane. Thus the sound reaching the sending diaphragm was electrically duplicated by the receiving diaphragm.

Gray filed the U.S. Patent Office's caveat, a formal notice of his claim to the idea of the new instrument, on February 14, 1876. He hoped that his caveat would prevent others from patenting the same idea within the period of a year. On the same day, but only a few hours earlier, Bell had applied for a patent for the same type of instrument. In later years there was a great deal of bitter legal dispute about priority, but in the end Bell was awarded the patent rights, and he received the credit for his invention.

BELL'S EXPERIMENTS

Bell was born in Scotland and educated in Edinburgh and London universities. At the age of twenty-three he emigrated with his father to Canada and settled two years later in Boston. Like his father and his grandfather, young Bell was then devoting his life to educating the deaf, and he had acquired a considerable knowledge of the physiology of human speech and hearing. For example, Bell hoped his deaf students could learn to speak better by comparing their attempts, displayed as visible speech, with a previously recorded pattern. However, designing this type of display was not easy with the instrumentation of the time.

Initially, he used both the phonoautograph and the manometric capsule as "voice oscilloscopes." The phonoautograph used a membrane to drive a quill that inscribed sound waves on moving smoked glass. The manometric device, invented by Koenig, voice modulated a gas flame viewed as a continuous wavering band of light in a revolving mirror. Unhappy with these processes, Bell built an improved version that used a human eardrum obtained from the Harvard Medical School.

In later years, Bell wrote that it had struck him how the delicate thin membrane of the human ear was capable of operating the very massive bones of the human ear. The thought occurred to him that a thicker and stouter piece of membrane should be able to move a piece of steel "and the telephone was conceived," he concluded.

During his experiments on June 2, 1875, one of his reeds got stuck to its electromagnet. When Bell told his assistant Thomas A. Watson to pluck the sticking reed, Bell found in the adjoining room that the corresponding reed began to vibrate and produced a sound of the same pitch. From this simple phenomenon, Bell deduced correctly that if a single sound could be transmitted electrically, so could complicated human speech and even music. A circular piece of gold beater's skin was stretched over a small cylinder into which one could speak, and the skin was connected to a reed over an electromagnet. After preliminary trials, the first complete sentence was spoken over the telephone: "Mr. Watson, come here, I want you." The date was March 10, 1876.

Although these early calling devices were crude and communications were poor, development was rapid in the U.S. Bell showed

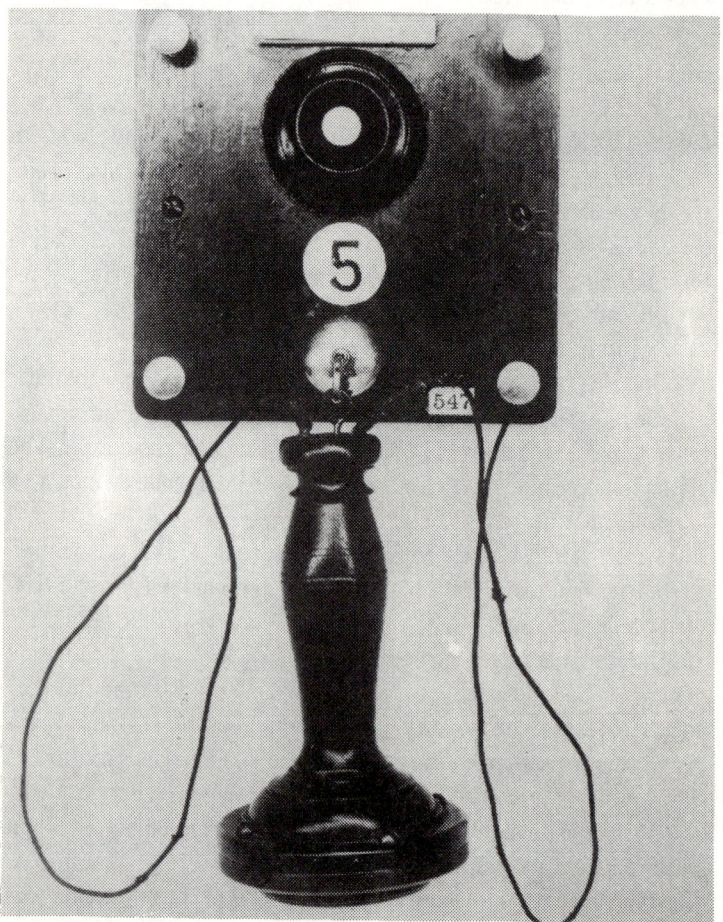

Fig. 1-1. First subscriber's telephone set of the type used at New Haven, Conn. (Courtesy of AT&T).

his equipment at the Philadelphia Centennial Exhibition of 1876, where the Emperor of Brazil, Dom Pedro de Alcantra, was very impressed. Sir William Thomson, later Lord Kelvin, was one of the judges at the Exhibition, and he wrote "With somewhat more advanced plans and more powerful apparatus, we may confidently expect that Mr. Bell will give us the means of making voice and spoken word audible through the electric wire to an ear hundreds of miles distant."

The first step came in the next year when an outdoor telephone line was run in Boston between the workshops of a Mr. Charles Williams, in which the first telephones were made by Watson, and Williams' private residence in Sommerville. Also in 1877, the first news dispatch was sent by telephone to the Boston *Globe*, and this feat inaugurated the public use of the telephone.

During his honeymoon visit to England in 1878, Bell was presented to Queen Victoria, and the telephone was demonstrated to her at Osborne, Isle of Wight. Communications were established with Cowes, Southampton and London. Bell did not spare himself in making his invention known. He lectured widely in the U.S. and first demonstrated the telephone in England to the annual meeting of the British Association at Plymouth in the autumn of 1877. Use of telephones expanded rapidly in the U.S. The first telephone and switchboard for commercial service was installed at New Haven, Connecticut, in 1878 (Fig. 1-1), with twenty-one subscribers.

Bell also offered his invention to Western Union, the leading telegraph company. Western Union, however, rejected Bell's offer to sell his telephone idea for $100,000. Marshall MacLuhan notes that in the minutes of Western Union's with Bell, they stated:

> The telephone is so named by its inventor A.G. Bell. He believes that one day they will be installed in every residence and place of business....Bell's profession is that of a voice teacher. Yet he claims to have discovered an instrument of great practical value in communication which has been overlooked by thousands of workers who have spent years in the field....Bell's proposals to place his instrument in almost every home and business place...is fantastic... The central exchange alone would represent a huge outlay in real estate and buildings, to say nothing of the electrical equipment....In conclusion, the committee feels that it must advise against any investments in Bell's scheme. We do not doubt that it will find users in special circumstances, but any development of the kind and scale which Bell so fondly imagines is utterly out of the question.

Fig. 1-2. An 1897 desk dial telephone (Courtesy GTE Automatic Electric).

Bell, however, smiled and continued to improve and publicize his telephone. In July of 1877, Bell and his financial backers, Sanders and Hubbard, formed a corporation, The Bell Telephone Company.

The first telephone line between Boston and Providence was opened in 1880; Boston to New York in 1885; New York to Chicago in 1892; New York to Denver in 1911; and New York to San Francisco in 1915. Figures 1-2 through 1-8 illustrate some of these early phones. Much of this early expansion came through the efforts of Theodore N. Vail, who had become, at the age of 33, the first general manager of The Bell Telephone Company. Despite stiff competition from other telephone systems devised by other inventors, including Thomas Edison and Elisha Gray, Vail promoted Bell business beyond the New England area for the next nine years.

From the beginning, Vail saw the telephone business as a public service. When he began in 1885 to connect local Bell companies with the long distance lines of his newly formed AT&T, American Telephone and Telegraph Company, he cooperated with the independent companies.

THE BELL SYSTEM

By 1976 the Bell System had over 118 million telephones in service. On an average business day, 470 million calls are made. In a year, over fifty million calls go overseas and can be dialed directly

Fig. 1-3. An 1892 dial telephone (Courtesy GTE Automatic Electric).

Fig. 1-4. An 1896 wall telephone with dial flanges (Courtesy GTE Automatic Electric).

from more than 200 U.S. cities and towns. Vail's goal has nearly been met. Americans can now reach 98% of the world's 350 million telephones.

The Bell System today is a group of companies that employs nearly one million people. The parent corporation, AT&T, owns the other companies in full or in part. The twenty-three operating telephone companies associated with AT&T operate in every state but Hawaii and Alaska and serve 80% of the nation's telephones. Service to the rest of the U.S. is handled by 1,700 independent companies which interconnect with Bell lines. These various independent companies, some of which serve several states, are linked by AT&T's Long Lines Department. Besides radio and television programs and computer data, Long Lines handles overseas service by radiotelephone, submarine cable, and satellite.

AT&T owns the Western Electric Company, which manufactures, supplies, and repairs equipment for the Bell System. Western

Fig. 1-5. Bell System's Subscribers telephone, CA. 1920 (Courtesy AT&T).

Fig. 1-6. Magneto wall set, 1907. Some of these wall phones are still in service in rural areas (Courtesy AT&T).

Electric was brought into the Bell System in 1882, just six years after the invention of the telephone. In support of its manufacturing operations, Western Electric performs a variety of service functions. Engineers tailor each major order for central office switching and transmission equipment to meet the specific service requirements designated by the telephone company for that order. Assuring that new equipment is compatible with the existing network is an essential part of their job.

In addition to providing its own manufactured products, Western Electric also purchases a wide variety of telecommunications and general products at the telephone companies' request. Western Electric stocks and distributes products of its own manufacture and purchased products through service centers located throughout the United States. These service centers also contain shops that repair and recondition used telephone equipment which is returned by the telephone companies, thereby conserving resources through reuse.

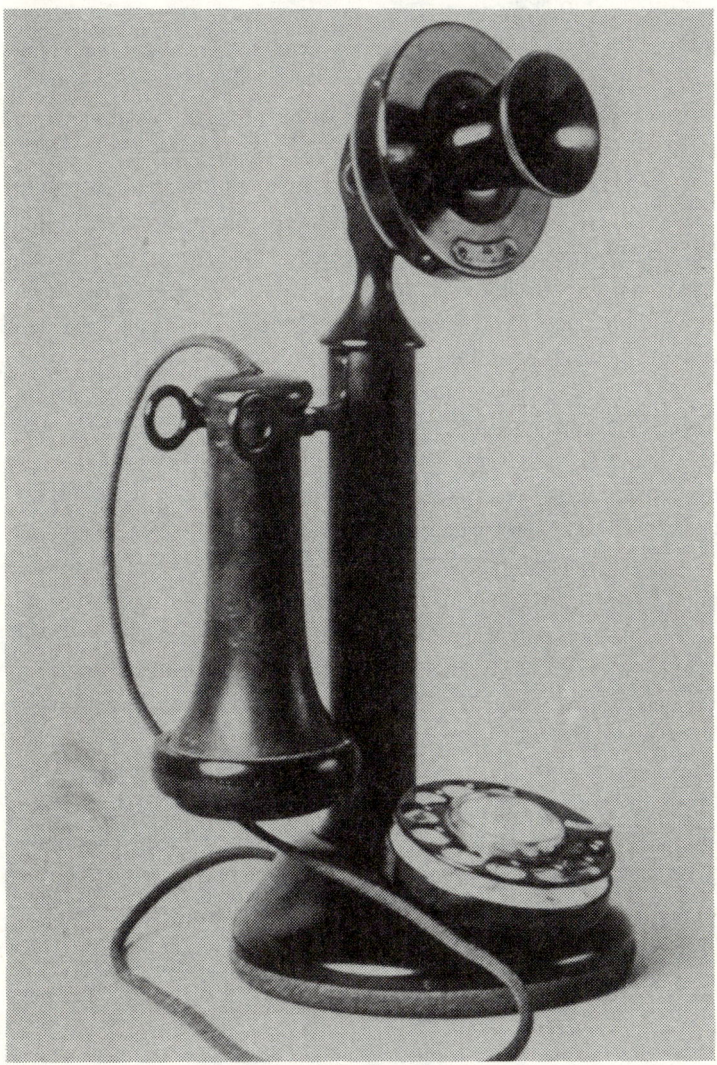

Fig. 1-7. Dial telephone, CA. 1919 (Courtesy AT&T).

AT&T owns another company, Bell Telephone Laboratories, a company devoted entirely to research and development for the Bell System. Research and development in the Bell System really began when Bell and Watson found ways to improve their telephone instruments a century ago. The actual Bell Laboratories, however, were created in 1925 to consolidate the research and development work for the Bell System. The numerous discoveries at Bell Labs since 1925 have included many breakthroughs leading to economical, efficient communications. The work of engineers and scientists at Bell Labs has averaged one U.S. patent for every working day since 1925, and two per working day in recent years.

Bell Labs transmitted the first pictures over telephone wires in 1924. C.J. Davisson demonstrated the wave nature of matter at Bell Labs in 1927. Karl G. Jansky developed radioastronomy there in 1933. The coaxial cable transmission system developed at Bell Labs in 1929 was put into use by 1936, and today can handle 108,000 conversations at one time.

One of Bell Labs' most remarkable achievements was the development of the transistor by John Bardeen, William Shockley and Walter H. Brattain in 1947. They began studying semiconductors, materials between insulators (like glass) which block the flow of electric current, and conductors (like copper), which let the current flow easily. Recalling the crystal "cat's whiskers" detectors in early radios, they rummaged through secondhand stores to find some. They reasoned that if a crystal, which is a semiconductor, could detect a radio wave, it might also be able to amplify that radio wave the same way a vacuum tube does by making a strong current vary according to the weak current of the signal being detected. One day they hooked up a germanium crystal with two wires just two-thousandths of an inch apart, amplifying a voice forty times. They had discovered the transistor effect. Today, new techniques have given us the integrated circuit in which thousands of transistors can be deposited and interconnected in an area less than a tenth of a square inch.

GTE AUTOMATIC ELECTRIC COMPANY

The story of GTE Automatic Electric began in 1889 in Kansas City, Missouri, when Almon Brown Strowger became irritated over

Fig. 1-8. Cradle type dial desk phone, CA. 1927 (Courtesy AT&T).

his telephone service. Convinced that he had lost customers because an operator had failed to complete a call for him, he vowed to devise a telephone system in which the calling party would directly control the equipment, completing connections without an operator.

Strowger's invention, which resulted in the dial telephone, marked an important chapter in the history of communications. Bell had invented the telephone itself, but had made no provision for interconnecting large numbers of phones. Early workers in the industry had devised simple switchboards to which telephones could be connected so that operators could interconnect any calling telephone to any other telephone at the request of the calling party. As telephones became more popular, switchboards became larger and the traffic heavier. Not only did the operators' duties become more complex, but it became apparent that there would be neither enough space for the switchboard equipment, nor a big enough supply of operators to handle mounting volume of telephone switching. It was many years, however, before switching equipment was sufficiently developed to permit dial installation in larger cities. New York City, for instance, began to get dial service in 1922.

Like so many great inventions, Strowger's idea was basically simple. Using electromagnets energized by electrical pulses received over the telephone line, his device actuated a pawl-and-ratchet mechanism that would move a metal finger (now called a wiper) over a bank of contacts, each connected to a different telephone. The calling telephone would be permanently connected to the wiper, and by sending the proper number of pulses, the caller would direct the wiper to the specific contact connected to the desired telephone. This process linked the calling and called telephones electrically. Additional pulses sent over another wire rang a bell at the called telephone at the opposite end of the line.

On March 12, 1889, Strowger filed his U.S. patent applications, and on March 10, 1891, U.S. patent #447,918 was issued. Strowger's next problem was to obtain financial backing for the production and marketing of his system. In 1893, he met an enterprising salesman, Joseph Harris. The two joined forces in the formation of the Strowger Automatic Telephone Exchange to develop and promote the new automatic system.

In 1901, the GTE Automatic Electric organization was formed, and acquired from the Strowger Automatic Telephone Exchange the U.S. rights to manufacture and sell Strowger equipment. Other acquisitions in the following years made GTE Automatic Electric the largest independent telephone company.

ITT: INTERNATIONAL TELEPHONE TELEGRAPH

ITT, incorporated in 1920, operates today in more than eighty countries around the world. Starting as a small telephone operating company in the Caribbean, initially as the owner of telephone operating companies in Puerto Rico and Cuba, the company was entrusted in 1924 with the modernization of Spain's telephone network. In 1925 ITT became truly international when it purchased the assets of the International Western Electric Company. Through that acquisition, ITT became a telecommunications equipment manufacturer with plants in Argentina, Australia, Belgium, Brazil, China, France,

Great Britain, Italy, Japan, Norway and Spain. Other companies were added later in Europe, Latin America, Africa and the U.S.

In addition to providing ITT with a strong manufacturing base, International Western Electric Company brought to the company not only research facilities, but also a group of international engineers and scientists who helped provide ITT with new technologies and transnational teamwork.

With the laying of the trans-Andean cable, ITT provided South America with its first intercountry telephone system and its first radiotelephone service to Europe and North America. ITT has provided Europe with its first multichannel commercial radiotelephone link, its longest coaxial cable network, and its first nationwide and international subscriber-to-subscriber telephone dialing system. A total of 103 countries on six continents now have 40 million local telephone lines and 2.3 million toll exchange lines provided by ITT.

While launching long-range research into such advanced projects as electronic switching, ITT's European engineers developed a highly successful switching system based on the crossbar concept, called the Pentaconta telephone switching system. In the 1950s this system was adopted in Europe and other continents around the world, followed by the Metaconta system which features stored-program-control techniques and reflects a major step into computerized switching technology.

ITT is bringing to the world markets digital switching, fiber optic communications, and a host of systems for total and flexible communications.

TODAY'S TELEPHONE

Advanced communications technology, assisted by transistorized electronics and computers, has made the telephone a versatile, fast and easy to use instrument.

In 1951 Direct Distance dialing began for U.S. calls, and in 1966 for calls overseas. Along with this convenience, the Bell System developed electronic switching systems, called ESS for short. It took ten years to put the first ESS in service in 1965, but by 1976 over 800 central offices had them, with many more to follow.

In 1963, Bell introduced the Touch-Tone (Registered Trademark of the Bell System) replacing the rotary dial with pushbuttons to make dialing simpler and faster. In the 1960's Dataphone (Registered Trademark of the Bell System) service enabled the telephone to function as another keyboard for the computer. Computer data, including diagrams, could be sent or received.

Today's telephone can do much more than just carry your voice. It can transmit TV programs, photographs, and computer data. Doctors can even read electrocardiograms sent over telephone lines. Presently, telephone companies are even experimenting with beams of light that can carry telephone signals just as radio waves do.

These developments are but a few discussed in this book that began some hundred years ago with Alexander Graham Bell's invention of the telephone.

How Your Telephone Works

The major objective in telephony is to provide voice communications over large distances. The traditional approach has been to convert the sound waves of your speech into electrical energy that can be transmitted efficiently over wires. The electrical waves, which ideally are identical in both form and frequency with the original sound waves, enter the receiver in the earpiece of your phone set and then are converted back to sound waves corresponding more or less to the original speech pattern.

In addition to enabling you to talk and listen, the phone set must be able both to select any other set in the system for a connection and to indicate to you when it is in use. Each set in a given geographical area is connected by way of a pair of copper wires called the *loop* to a central office that is a switching center. Phone calls within the area are routed through that office; calls to numbers outside the area are routed to other offices through interoffice trunk lines.

Picture yourself initiating a phone call. You lift the handset of your telephone (Fig. 2-1A), whereupon the central office returns a dial tone to let you know that a line is available. You then dial a number. The central office selects a correct pair of wires and sends a ringing signal to set B (Fig. 2-1B). If someone answers set B by lifting the handset, the central office connects you to set B. Communication is thereby established. The central office also senses termination of the call and releases the lines for other calls.

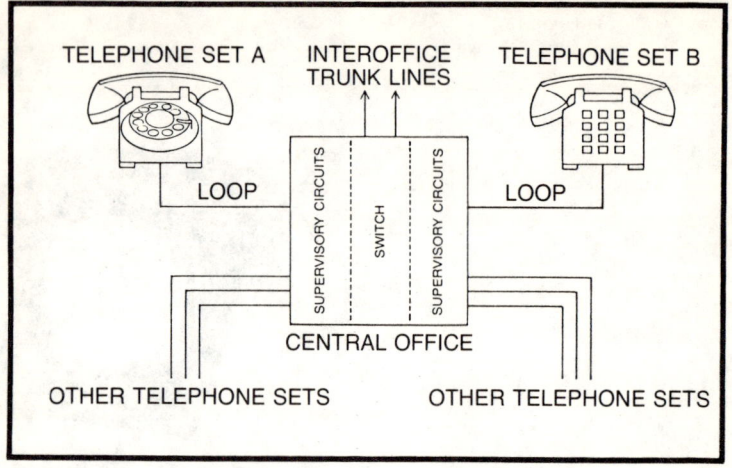

Fig. 2-1. Basic functions required to make a phone call.

THE LOOP

The central office furnishes the current for operation of your telephone. The standard voltage most widely used is 50V DC. If you measure the voltage at your phone set with the phone on the hook, 50V is what you expect. This voltage is placed on your line in such a way that no shock or fire hazard is created. Even if you short-circuit your line by touching the wires together, only about 1/10 of an amp will flow through your body.

Your line consists of a pair of relatively small wires (Fig. 2-2A). Therefore, the farther you are located from the central office, the higher the resistance of the line becomes. Since the resistance of the line increases with distance, the amount of current that can be delivered to your phone decreases as your distance from the central office increases. Hence, there is a limit as to how far you can be located from the central office, and this limit is generally determined by the "loop resistance" of the pair of wires in Fig. 2-2A that serves your set. If your phone is very close to the central office, the current in your line may be as high as 0.08 of an amp when the phone is off hook. If your phone is many miles away, the current may be as small as 0.02 of an amp. Your telephone is a "current operated device," and it cannot work without sufficient current supplied by the central office.

THE TELEPHONE SET

A typical telephone set consists of a handset (Fig. 2-2B), containing a transmitter and receiver, that is connected to the base with an extendible cord (Fig. 2-2C) holding four conductors. The telephone houses a transformer system (Fig. 2-2D) designed to separate outgoing and incoming speech signals, a switch to connect and disconnect the phone and the loop (Fig. 2-2E), a dial pulser (Fig. 2-2F) to signal the required numbers to the central office, and a bell (Fig. 2-2G) to signal incoming calls.

The Handset

The handset (Fig. 2-2B) of the typical telephone contains a transmitter for relaying electrical signals and a receiver to convert the electrical signal back into voice tones.

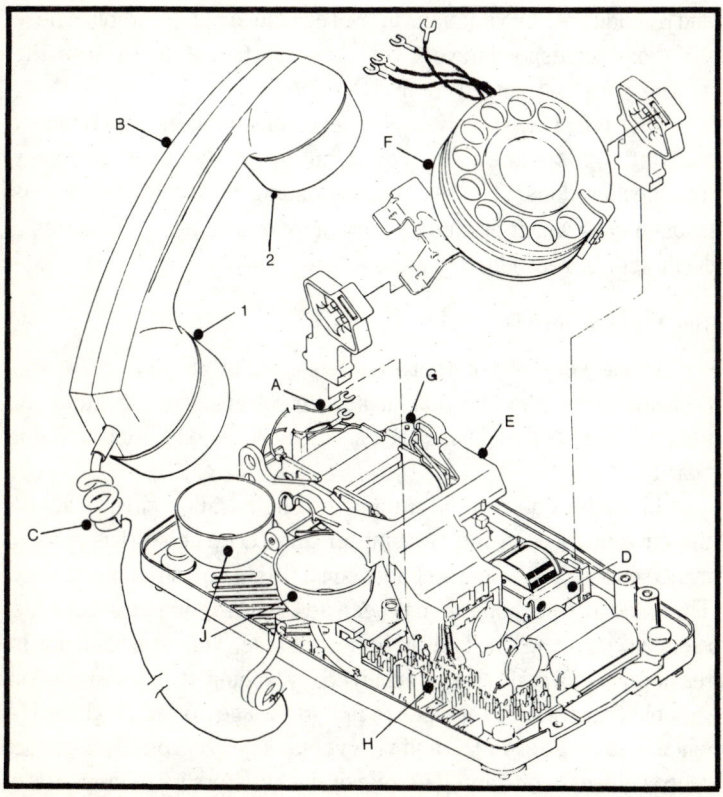

Fig. 2-2. Parts of a telephone (Courtesy of GTE Electric Automatic).

When you lift the handset of your phone, a circuit is completed and a current starts to flow which signals the central office that you want service. This current has one important function: it powers the transmitter in your handset. This transmitter, actually a carbon microphone, produces an AC electrical signal when you speak into it. However, this particular microphone is designed to work only if it has a relatively high DC flowing through it first.

The carbon microphone works by varying the resistance of a loosely packed group of carbon granules. A sensitive diaphragm responds to your voice and alternately compresses and expands the granules. When the carbon is compressed, its resistance goes down and the DC through the carbon increases. When the carbon expands, the resistance goes up and the current drops. Thus, the DC supplied by the central office is fed through the carbon microphone and is modulated by variations in the resistance of the carbon. These variations correspond to your voice, so a portion of the DC from the central office is changed to an AC voice current.

The telephone receiver is a very sensitive, small permanent magnet loudspeaker. It works the same way an ordinary loudspeaker does. The AC corresponding to the voice, passes through coils which produce a varying magnetic field and causes a diaphragm to reproduce the voice.

The Transformer

Inside your phone there is an assembly of coils, resistors, capacitors, and the like that makes up a transformer system, or network (Fig. 2-2D). These are usually mounted on a printed circuit board.

In modern networks, automatic compensation circuits adjust the current through the transmitter and receiver so that performance on a very long line is almost equal to that on a short phone line. The standard telephone set has been designed to compensate for the reduced current in long lines in some measure. This compensation is termed *equalization* and is achieved by adding two varistors, or variable resistors, to the speech network. They offer a high resistance to low values of DC and vice versa. The DC voltage from the central office is constant. The role of the varistors is to compensate for the fact that in short loops the current is relatively high, tending to make voices too loud, and that in long loops the current is relatively low, tending to force talkers to raise their voice.

Circuits in the network divide and combine incoming and outgoing speech signals in the proper proportions. In a properly designed telephone, a small portion of the AC speech from the transmitter is fed into the receiver to create what is called *sidetone*. This sidetone allows you to hear yourself when you talk. However, the level of sidetone is carefully controlled. If all the transmitter current were passed through your own receiver, you would sound so loud that you would tend to whisper and the party would not be able to hear you. On the other hand, if there were no sidetone, you would tend to shout.

The network printed circuit board (card) provides terminals (Fig. 2-2H) so that all the telephone components (ringer, handset, dial, etc.) can be connected together.

The Hookswitch

The hookswitch (Fig. 2-2E) consists of a group of normally open or normally closed contacts that switch circuits inside the phone. When the handset is lifted, the speech network is connected to the loop, causing DC to flow. The current actuates a relay in the central office, and the relay connects the loop to the supervisory circuits. The central office returns an audible tone, indicating to you, the caller, that the system is ready to receive the number of the person you want to reach. On some phones, the hookswitch is mounted on the network, but, on most phones, it is a separate assembly. The plunger of your phone actuates the hookswitch, usually via a lever arrangement.

The Ringer

A telephone ringer consists of a single gong (Fig. 2-3A) or two gongs (Fig. 2-3B) which are struck by a clapper. AC through the coils of the ringer causes a rapidly fluctuating magnetic field to be produced. This magnetic field, in turn, causes an armature to move back and forth. A small permanent magnet is used in the assembly to improve its sensitivity. A phone ringer is always used in series with a capacitor so that AC current can flow through the ringer, but DC current cannot.

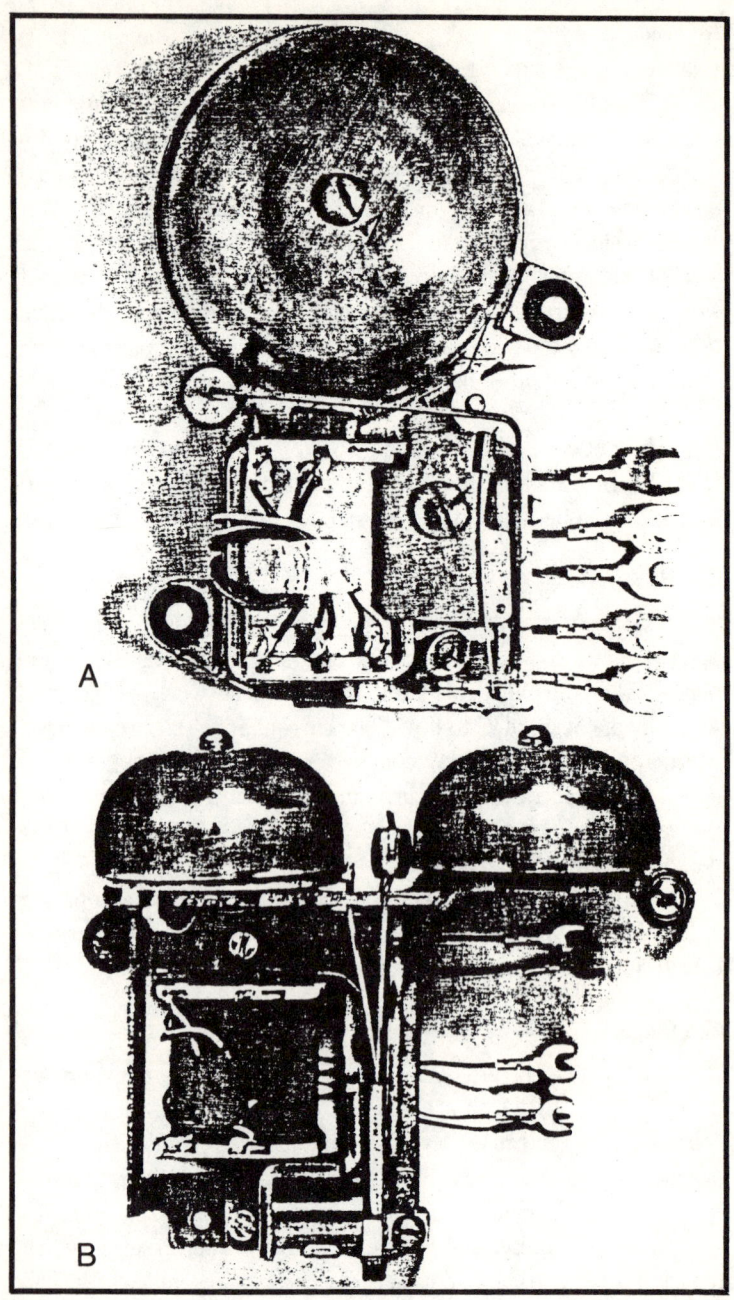

Fig. 2-3. Types of bells for ringing called lines (Courtesy of ITT).

The single gong ringer is a single coil tapped at 500Ω, 1000Ω, and 2650Ω to provide any required resistance to ground for tip party identification. Figure 2-9B shows the two gong ringer, whose coil permits either 1000Ω or 2650Ω winding. Below are the ringer frequencies:

Harmonic	33⅓ Hz
	50 Hz
	66 ⅔ Hz
	16 ⅔ Hz
	25 Hz
Synchromonic	30 Hz
	42 Hz
	54 Hz
	66 Hz
	16 Hz
Decimonic	20 Hz
	60 Hz
	30 Hz
	40 Hz
	50 Hz

The Dial Pulser

The rotary dial (Fig. 2-4) is a rather complex assembly which is used to rapidly open and close the line circuit to signal the central office and transmit the number of the called party.

When you turn the dial, a spring is wound up, and an auxiliary set of contacts, called *shunt springs* are actuated. These shunt springs switch some circuits so that the pulsing will not be heard in the receiver. When you release the dial, a ratchet mechanism engages and a cam rotates to open and close contacts, and pulse the line to the central office. The return of your dial is controlled by a govenor so that only six break pulses are conveyed to the central office at the correct rate of speed, ten pulses per second, with an interruption of approximately a sixteenth of a second between each pulse. During pulsing, the contacts are open for a longer period of time than closed.

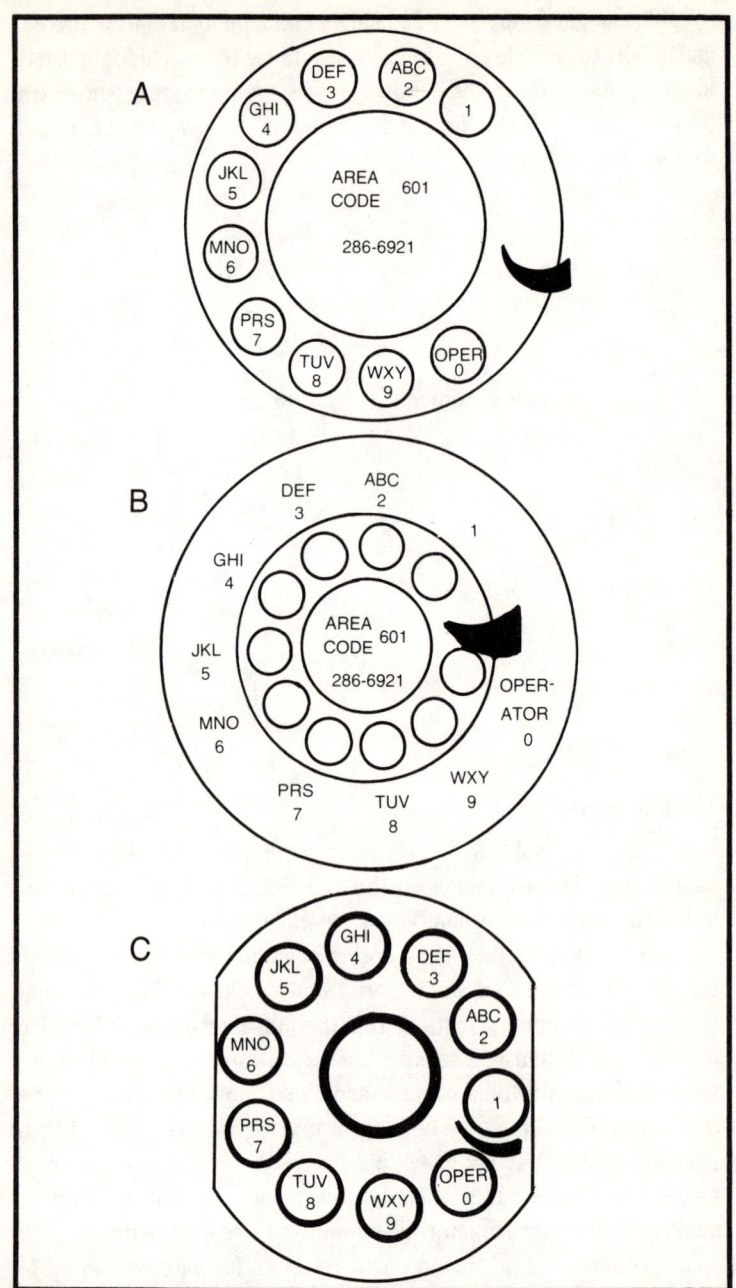

Fig. 2-4. Rotary dial pulsers (Courtesy of ITT).

The ratio of open time to closed time is referred to as *percent break*, and the usual requirement is for the dial to produce a pulse of approximately 61% break. So, if you dial a "2," the central office will "see" and open a period of 0.061 seconds, followed by a closed period of 0.039 seconds (= 100%), then another opening of 0.061 seconds. The line will then close and stay closed until another digit is dialed.

Rotary Dials are submitted to rigid quality control to insure smooth; quiet and trouble-free operation. Each dial consists of a rigid metal base upon which are mounted the gear train, mainspring, contact spring assembly, numeral ring, clear plastic finger plate, and station number card. The gear train and contact spring assembly are protected by a plastic dust cover. Leads are equipped with cord tips designed for #4, #5, or #6 screws. Dials are adjusted to a speed of ten dial impulses per second nominal, and a pulse ratio with a break period of 61.5% of the pulse duration. A governor minimizes the forcing of dial return. Regular dials have numerals only; metropolitan dials have letters and numerals (Fig. 2-4A). Dials coded H have only a dot at each finger hole (Fig. 2-4B). Figure 2-4C shows a miniature dial for use with dial-in-handset telephones.

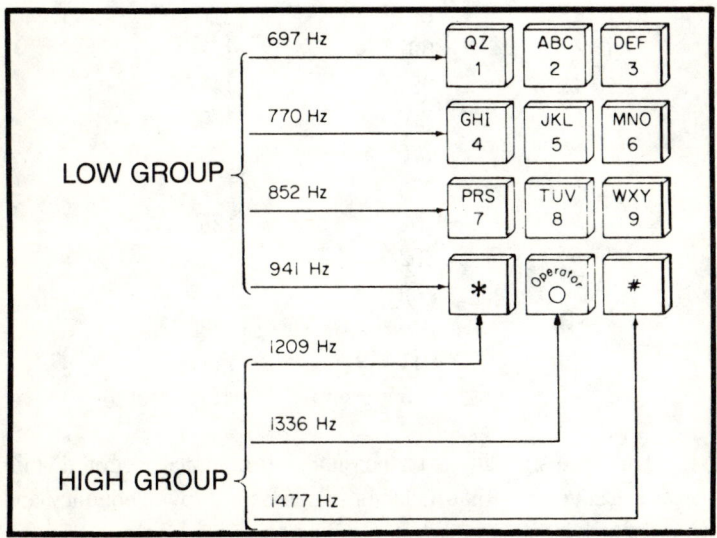

Fig. 2-5. All digits are derived by combining one low group and one high group frequency.

Touch-Tone[R] is a signaling system that is rapidly replacing the rotary dial system. Instead of sending groups of pulses to the central office, as with the rotary dial, Touch-Tone sends tones which are deciphered by special receiving components in the central office. Each time you depress a button on your Touch Calling Unit (TCU), a pair of very accurately controlled tones are produced and transmitted to the central office. Pairs of tones are used so that the detector in the central office will accept the signal only if both tones are present simultaneously. Because tones are in the voice frequency range, dual tones are used so that the central office can distinguish the difference between actual signaling and voice noise.

Frequencies are grouped as follows:

697.770, 852 and 941 Hz (Herts) = low group
1209, 1336 and 147 = high group

Digit	Tone	Frequencies
0	941 and	1336 Hz
1	697	1209
2	697	1336
3	697	1477
4	770	1209
5	770	1336
6	770	1477
7	852	1209
8	852	1336
9	852	1477
*	941	1209
#	941	1477
	(Low group)	(High group)

This tone signaling is based on an international accepted standard of frequencies that includes four tones of low frequency and three of high frequency. When you push the 5 button (Fig. 2-5), the high group tone is 1336 Hz (cycles) per second, and the low group tone is 770 Hz. These tone frequencies must be accurate within

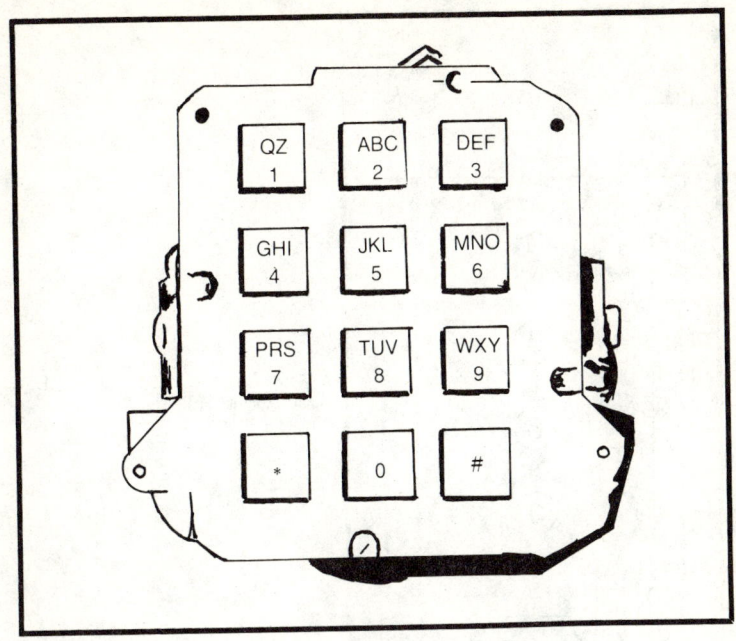

Fig. 2-6. Pushbutton Pulse-Touch dial (Courtesy of ITT).

1.5% of their stipulated value under all conditions of temperature throughout the life of your set. It achieves this level of accuracy by a method of digital division. (See Chapter on Analog vs. Digital Communications).

Figure 2-6 shows a Pushbutton Tone Dial, which emits impulses similar to a rotary dial. It provides the convenience of pushbutton dialing in areas where tone dialing can not be accommodated by the central office.

Figure 2-7 shows the Pushbutton Tone Dial used in areas where the central office can handle Touch-Tone dialing. It has a twelve button matrix and highly reliable transistorized oscillators. Ten buttons are used for dialing and two special service buttons are included to avoid future obsolescence.

THE SWITCHING NETWORK

The basic function of a telephone exchange, or central office, is to connect you with another subscriber in a network. You, as the caller, are connected through a control mechanism to a called line. In

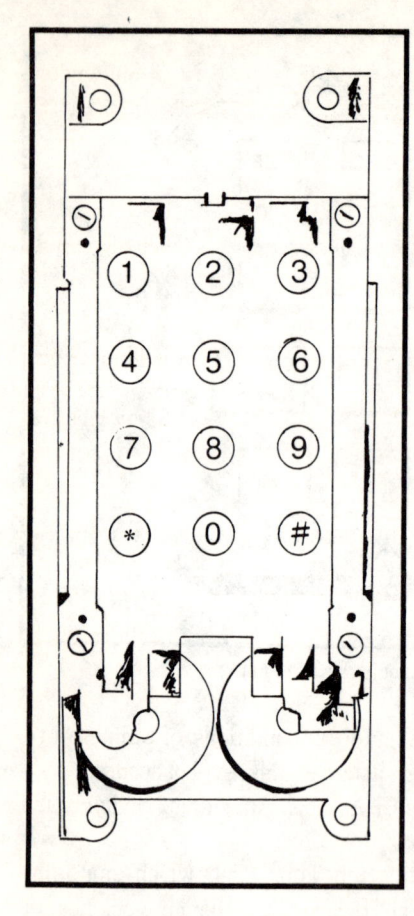

Fig. 2-7. Pushbutton Tone Dial (Courtesy of ITT).

early exchanges, the control was an operator who manually connected parties. Relatively early in the history of the telephone, the control became automatic (Figs. 2-8, 2-9).

Types of Exchanges

Exchanges may be classified by the type of network in which they operate. The PAX is a private, automatic exchange, interconnecting subscribers in a private telephone network, but not having access to subscribers in the public network. The PABX is a private, automatic branch exchange which interconnects subscribers in a private telephone network, as well as allowing access to subscribers in the public network.

Public Local Exchanges are operated by the public communication operating companies, such as AT&T, GTE and Independent Telephone Companies, to which the general public, you and I, are subscribers. These exchanges generally cover a relatively small, local geographical area.

Toll Exchanges are those that the public operating companies use to interconnect the subscribers in different local exchange areas on a nationwide and worldwide basis. Connections between toll exchanges are made over *trunk* or *toll* lines.

There are several basic parts to a telephone exchange. Generally, your access to the exchange is through a line circuit, providing the interface between the exchange equipment and your distribution plant. The line circuit also assists in detecting a request for service when you lift the handset from the hookswitch.

In many systems, the line circuit connects the calling line to a *register* that returns dial tone and accepts dial pulses.

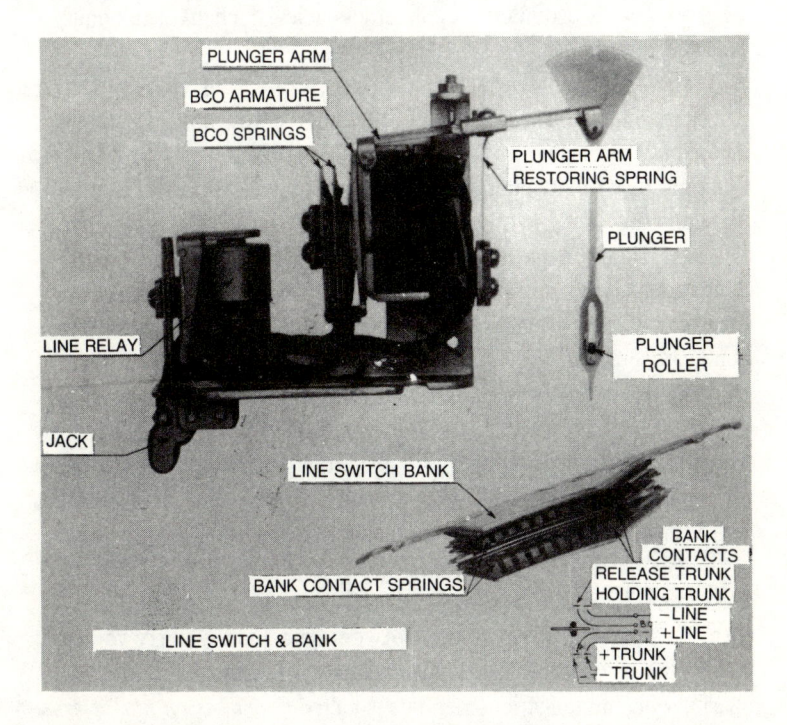

Fig. 2-8. Old line switch, CA. 1901 (Courtesy of G T E Automatic Electric).

The Exchange Control

The *control* is the brains of the exchange. This control interprets the dialed number and instructs the *switching matrix* circuits how to connect the calling and called lines. This control also performs busy tests on called lines and rings called lines.

In early automatic telephone exchanges, the control was an assembly of electromechanical relays wired in a fixed pattern. This type of control is called *wired logic* control. The assembly of relays is called a *register* or a *marker*, depending upon the system.

In modern electronic exchanges, control is accomplished through an SPC (Stored Program Controlled) special purpose digital computer. This type of control offers greater flexibility to the exchange than was possible with the wired logic controls. An SPC provides the customer such features as conferencing of calls, call forwarding to predetermined locations, number tracing, routine traffic analysis, and automatic billing. Generally such features cannot be economically provided in electromechanical exchanges not equipped with an SPC.

The connection between your circuit and your party's within the exchange is accomplished by selecting a path among a multiplicity of paths available in the exchange. In early electromechanical systems, each individual call occupied its own path and the many simultaneous calls in the exchange were handled by providing a different path for each call. The multiple path circuitry is called *space division switching*, and the paths are arranged by controlled opening or closure of a series of relays such as crossbar switches or semiconductor diodes (PNPN diodes) arranged in a switching matrix. This control is effected either by wired logic or stored program control.

More recently, increasing availability of high technology equipment has made it possible to handle the many simultaneous calls in an exchange in a different manner. There are no fixed position switching matrixes in this *digital switching system*. All conversations are sampled in different time intervals, converted to pulses, and in turn, these pulses are switched between you and the party you call over communication channels in exact fixed time intervals. The time sharing of communication channels is called *time division multiplexing*, or TDM. If the pulse samples change in number and arrange-

Fig. 2-9. Central office selectors as they were used in 1897 (Courtesy of GTE Automatic Electric).

ment according to a predetermined code corresponding to the amplitude of the original wave, the system is of the pulse-code-modulation, or PCM type.

Each exchange has a size within which it is most efficient, ranging from the maximum number of lines that can be connected, down to a minimum number below which the cost per line becomes uneconomical.

HOOKING UP YOUR OWN PHONE

If you already have an extension phone in your house that plugs into a wall jack, your problems connecting your phone will be minimal. Simply unplug the phone company's instrument and plug it in. In most instances, though, this is not the case.

A source of phone connectors and the like are readily available through many companies including Saxton Products and Radio Shack. These companies manufacture all types of plugs, sockets, and extensions for telephone equipment.

The following paragraphs will provide you with some sound advice when attempting to install your own equipment.

Let's say you want to extend your phone service into another room with a 25-foot extension. If there is no removable plug at the present location, it'll be necessary to go directly to the terminal block located on a baseboard near the phone.

You need what is termed an instant jack of the type available through Saxton Products and Radio Shack. Once you have located the terminal block, remove the center cover screw to gain access to the four wire terminals. There is a keyway located on the phone company's terminal block and your instant jack. Simply place the instant jack over the terminal block, then insert the center screw in the hole provided in the instant jack.

With the instant jack in place it is a simple matter to run an extension line to another room. Extension lines available plug directly into the instant jack. Use insulated wire tacks to run the extension line along the baseboard.

Generally, when connecting any phone with the Class A style 4-prong plug, you simply remove the screw holding the plug together. Then notice that two of the prongs are closer together than are the other two. The other two prongs with the greatest separation are marked R and G. Now, when you look at the phone wires, one will be red while another will be green. Connect the green wire to the G terminal and the red wire to the R terminal. The other two wires are not normally needed unless you have, for example, a lighted phone.

Once you have connected the red and green wires, place the holding screw back into the plug. Plug the phone into the jack to try it. If the phone fails to operate, try reversing the red and green wires; sometimes the leads are transposed inside the phone itself.

In general the most important point to remember when connecting phones, especially when you don't know what you are doing, is to *maintain the same color code* at the terminal block. A good idea is to always keep one phone in the house operating while hooking up another. That way you can always refer to the other if something goes amiss.

Analog vs. Digital Transmissions

In order to better understand the descriptions of the "smart phones" in the chapters that follow, a basic knowledge about transmission modes is necessary. The telephone channel reaching your home today is an *analog* channel, meaning that a continuous stream of frequencies is transmitted. Sound audible to humans consists of a spread of frequencies from about 30 Hz to 20,000 Hz (hertz). We cannot "hear" sounds below 30 Hz or above 20,000 Hz. Transmitting high-fidelity music along the phone lines into your home is technically possible, but would mean sending a continuous range of frequencies from 30 Hz to 20,000 Hz. However, for cost reasons, the central office transmits a range of frequencies that varies only from 300 Hz to 3400 Hz, which is enough to make your voice recognizable and intelligible but definitely not "high-fidelity."

These telephone signals travel over lengthy channels packed together, or *multiplexed*, so that one channel can carry as many signals as possible. To accomplish this, your voice might have been raised in frequency from the 300Hz-3400 Hz range to the 60,300 Hz-63,400 Hz range, and your neighbor's voice to the 64,300 Hz-67,400 Hz range. This way many signals can travel together on the same channel without interfering with one another, as long as the frequency range for each voice is different. The multiplexed "voices" are still transmitted in an analog manner, that is, as a continuous wave of frequencies (Fig. 3-1).

Fig. 3-1. An analog signal.

Almost all of the world's telephone companies grew up using analog transmission, and because of the enormous amount of money invested, this will remain so for the years to come. However, digital transmission is rapidly evolving, and major advantages in digital transmission are beginning to emerge. I'm sure that if the telecommunication companies were to start afresh, there would be almost entirely digital transmission, with the exception of local loops between a subscriber and his nearest switching office. These new telephone offices would use a technique called *pulse code modulation*, or PCM, in which the voice and other analog signals would be converted into a stream of *bits*, or pulses looking like computer data. *Digital* transmission means, then, that a stream of on-and-off pulses are sent, in the way data travels in computer circuits (Fig. 3-2).

A major advantage in using digital techniques for transmission is the cleaning up of line noise. In analog transmission, whenever the signal is amplified, the noise and distortion is amplified with it. As the signal passes through its many amplifying stations, the noise is amplified and is thus cumulative. With digital transmission, however, each repeater station generates the pulses. New, clean pulses are reconstructed and sent on to the next repeater, where another cleaning up process takes place. Therefore the pulse train can travel through a dispersive, noisy medium, but instead of becoming more and more distorted, it is repeatedly reconstructed, and thus remains unaffected by the corrosion of the medium.

If digital bit streams formed the basis of our communication links, then computer data would no longer need to be converted into

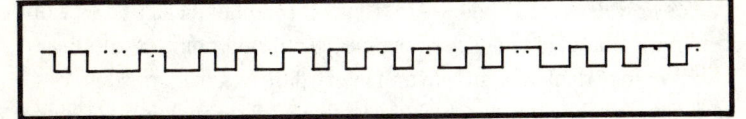

Fig. 3-2. A digital signal.

an analog form for transmission, as it is today. However, analog information such as the sound of the human voice needs to be coded in some way so that it can be transmitted in the form of pulses and then decoded at the other end to reconstitute the voice sounds. This is already done on many short distance telephone trunks. Converting bit streams into an analog signal is done through a device called a *modem*.

Transmitting data signals over systems that are digital from one end to the other avoids the use of digital-to-analog conversions by modems which are costly and distort the information content. Although the telephone network was developed originally as an analog network for voice transmissions, in recent years that network has been used increasingly for nonvoice information, such as computer data.

The Bell System has developed the DDS, Digital Data System, for office to office transmission of business information, packed in high speed streams of pulses by means of the Data-Phone^R Digital Service. The system carries information in the digital form common to most data processing equipment now in use. This enables customers to replace the modem data sets which are necessary today with much simpler and less costly digital station equipment.

An important long-term advantage to the digital transmission of the voice is the fact that all signals become a stream of similar looking pulses. Consequently, they will not interfere with one another and will not make differing demands on the engineering of the channels. In an analog signal format such as we have with most telephone lines, television and data are much more demanding in the fidelity of transmission than speech, and thus create more interference when transmitted with other signals. Eventually, perhaps, there will be an integrated network in which all signals travel together digitally.

LIGHTWAVE COMMUNICATIONS

In addition to using light just to see, telephone researchers are thinking of using light waves for the same purpose telephone companies now use radio waves—to transmit telephone calls, television programs and data messages digitally from one point to another.

We may think of light waves as very different from radio waves, but actually, they are much alike. Both are different forms of electromagnetic radiation. They differ only in the rate at which they vibrate or oscillate, that is, their frequencies. Light waves vibrate faster than radio waves—a thousand to a million times faster than microwaves.

Electromagnetic waves include all forms of radiation: visible light, infrared, ultraviolet, and X-ray, as well as radio waves and TV signals. Light waves constitute only a tiny part of all the electromagnetic waves in the universe. So far it has been possible to use only the lower frequency vibrations of the spectrum. Researchers are exploring the possibilities, however, of opening up the unused part of the spectrum, the infrared, visible light, and ultraviolet regions.

There unused regions beyond microwaves offer staggering possibilities for communications because, as the frequency of electromagnetic radiation is increased, so is the potential for carrying information. A beam of laser light waves has the potential of providing a million times more "space" to carry messages than the entire range of frequencies now used for telephone communications.

The Laser

Light is made by excited atoms. In an electric lamp, the passage of electric current makes a wire hot, exciting the atoms by energizing them. In a fraction of a second, each excited atom unloads the extra energy as a pulse of light. Each atom behaves as if the others did not exist, radiating independently in any direction. Also, the several different kinds of atoms in the wire radiate at different frequencies. As a result, the wave output from the wire is disordered.

In contrast, the laser performs the feat of organizing atoms into teams to produce ordered light. Such light is called *coherent*. Now, imagine a hollow glass cylinder with a mirror at each end. The cylinder contains atoms, all of which give off, when excited, waves of a single frequency. Suppose, too, we have a way to maintain most of the atoms in an excited state. As in an electric lamp, the atoms emit at random in every direction. But the radiation that moves along the axis of the cylinder is bounced back and forth between the mirrors.

Each time the radiation hits an excited atom, it stimulates the atom to emit its stored energy as a new pulse of waves which at once fall into step. Racing back and forth along the tube, the radiation builds up into an intense, single frequency beam of coherent waves. One end of the mirrors is made slightly transparent, allowing the beam to escape into the open.

Lasers especially suited for communications need to be developed. Here crystals offer exciting possibilities. Bell Labs' scientists are experimenting with a laser made from a crystal of yttrium aluminum garnet doped with neodynium. Known as the YAG laser, it generates a powerful, continuous beam at a frequency that is excellent for communications.

Along with suitable laser generators, there must be a means for modulating the beam, converting messages into light waves. For this, the laser beam must be carved into pulses. Furthermore, to realize the laser's giant capacity, the pulses must follow each other in close procession, yet remain distinct from each other. This has been accomplished by combining the YAG laser with another special crystal, allowing a parade of short pulses, each of which begins and ends within *thirty millionths of one millionth* of a second. So closely are the pulses spaced that *thirty billion* of them can be squeezed into one second, sufficient time to transmit, in pulse code, 330 TV programs at the same time.

Fiber Optics

Bell System's work on lightwave communications has involved a broad effort in many disciplines. Bell Labs' engineers and scientists operated the first semiconductor solid-state laser continuously at room temperature in 1970, and pioneered in developing a light-emitting diode specially designed for use with glass fiber lightguides.

An optical fiber is an extremely thin strand composed of two layers of different kinds of glass. Thinner than a human hair, an optical fiber has the ability to conduct light, much as a waveguide conducts microwaves, making it bend around corners. Simple in theory, the creation of efficient optical fibers depends upon controlling the composition, purity, and uniformity of microscopic glass layers to unprecedented precision. Figure 3-3 shows the designs of a fiber optic cable.

Since May of 1977, the Bell System has had under evaluation a full-service lightwave communication system in the city of Chicago. The system runs about 1.5 miles under downtown Chicago streets,

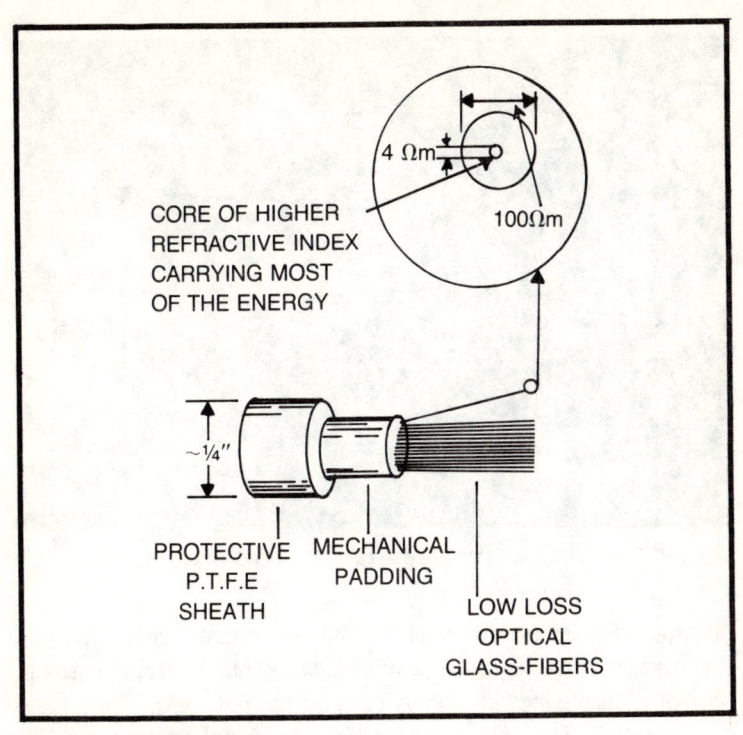

[Figure: Fiber optic waveguide diagram with labels: "4 Ωm", "100Ωm", "CORE OF HIGHER REFRACTIVE INDEX CARRYING MOST OF THE ENERGY", "~¼"", "PROTECTIVE P.T.F.E SHEATH", "MECHANICAL PADDING", "LOW LOSS OPTICAL GLASS-FIBERS"]

Fig. 3-3. Fiber optic waveguide.

and carries voice, data, and video signals and pulses of light transmitted through the glass fibers. Figure 3-4 is a photograph of the cable designed especially for the system. It is composed of twenty-four hair-thin lightguides arranged in two ribbons within the cable.

Figure 3-5 is a diagram of the Chicago lightwave communication system. The lightguide cable carries voice, data and video signals between the Brunswick building and the Illinois Bell Franklin central office, then between the Franklin office and the Wabash central office. The video signals originate from Bell System Picturephone[R] Meeting Service rooms. Not a single customer phone call has been interrupted by the lightwave components. The lightwave system has surpassed the 0.02% outage rate, with a projected annual outage rate of 0.0001%, or thirty seconds a year. On the average, less than one second per day has contained a transmission error, or, the system is 99.999% error free. Besides efficiently handling voice, data, and video communications, the system marks the successful

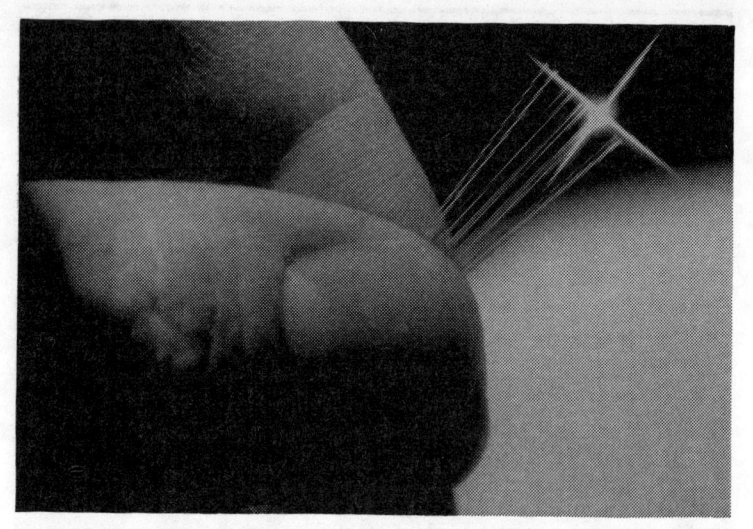

Fig. 3-4. The Bell Systems' fiber optic cable designed for use in Chicago.

incorporation of this new technology into both trunk facilities (part of the network connecting phone company switching centers) and loop facilities (linking individual customers to the network).

The lightguide cable used in Chicago is manufactured by Western Electric. The cable, only one-half inch in diameter, is a fraction of the size of cables for lower capacity systems now used to interconnect central offices in cities. A single pair of lightguides in the cable can carry 672 simultaneous conversations (at a 44.7 megabit per second rate), or an equivalent mix of voice and various types of data signals. Although this capacity was judged sufficient for the volume of traffic carried, lightguide cables with 144 fibers and a capacity of nearly 50,000 simultaneous conversations have been constructed and tested.

Each lightguide in the Chicago system is connected at one end to a transmitter module containing either a solid-state laser or a light-emitting diode (LED) light source, both smaller than grains of salt. The laser and LED transmitter are interchangeable for the Chicago application.

The other end of the lightguide is connected to a receiver module containing a tiny photodetector, known as an *avalanche photodiode,* that converts light pulses into electric signals. Terminal circuits then convert these signals into a format compatible with the nationwide telecommunications network.

Rapid interconnection of lightguide cables with little "leakage" of light was accomplished by a mass splice technique that permits simultaneous joining of all fibers with no handling of individual ones.

And the research goes on. Work continues at Bell Labs and elsewhere in many areas of lightwave communications. Theoretical and experimental work also continues on switching light pulses. Optical switches may someday be used instead of electromechanical and electronic switches to connect more calls at faster speeds than can now be done.

SATELLITE COMMUNICATIONS

Digital technology has also been responsible for improved telephone communications via satellite. The first commercially available digital satellite service for voice, data, and facsimile transmission was introduced last year by American Satellite, a domestic satellite communications company. ASC demonstrated simultaneous transmission and reception of digitized voice and high-speed data over a 56 kbps (56,000 bits-pulses-per second) live satellite channel. This

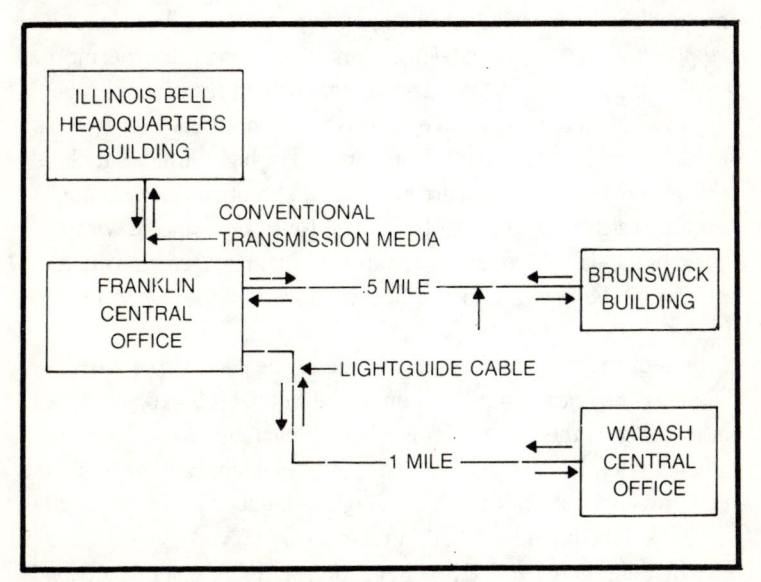

Fig. 3-5. The Chicago light wave communication system.

new ASC equipment eliminates the effects of satellite transmisison delay. For example, a Satellite Delay Compensation Unit (Fig. 3-6A) can increase the data throughput by two to three times or more in satellite channels, and can also increase the data throughput in terrestial channels.

Satellites that can serve as towers for telephone service are of two types: passive and active. *Passive* satellites are large balloons or other types of reflectors that simply "mirror" a signal without adding to its power. NASA's ECHO I sphere is a passive satellite. ECHO is simply a reflector. It acts as a backboard. Signals from the ground are bounced off its reflective surface to another location on the ground. Working with NASA, Bell Telephone scientists first bounced coast-to-coast telephone calls off ECHO I in August 1960.

Such satellites are relatively cheap and uncomplicated. But lack of amplification in the satellite means that even with expensive, high-power transmitters and very sensitive receiving equipment on the ground, the communications capacity and bandwidth are severely limited.

Active satellites contain amplifiers, similar to the amplifiers in overland microwave towers. Signals are received in the satellite from a ground transmitter. The signals are amplified by the satellite, then sent on to a ground receiving station overseas. More circuits are possible with active satellites than passive ones. Furthermore, smaller and less expensive ground stations can transmit signals.

In addition to being active or passive, communication satellites can also be high altitude or medium range. The high altitude satellites which have been proposed are *synchronous*, that is, their altitude is carefully calculated to make their speed in orbit coincide with the speed of the earth's rotation. This places them in a fixed position in relation to the earth, and the necessary altitude is about 22,000 miles.

Medium range satellites do not need to be placed in such orbits. The most practical altitudes seem to be between 5,000 and 8,000 miles. Each of the two systems offers advantages over the other, but Bell scientists are convinced that the medium range altitude is more practical at this time, although a number of high altitude satellites have been launched in recent years.

In Septemter of 1977 a new Intelsat IV-A communications satellite, developed by Hughes Aircraft, was launched by NASA. It

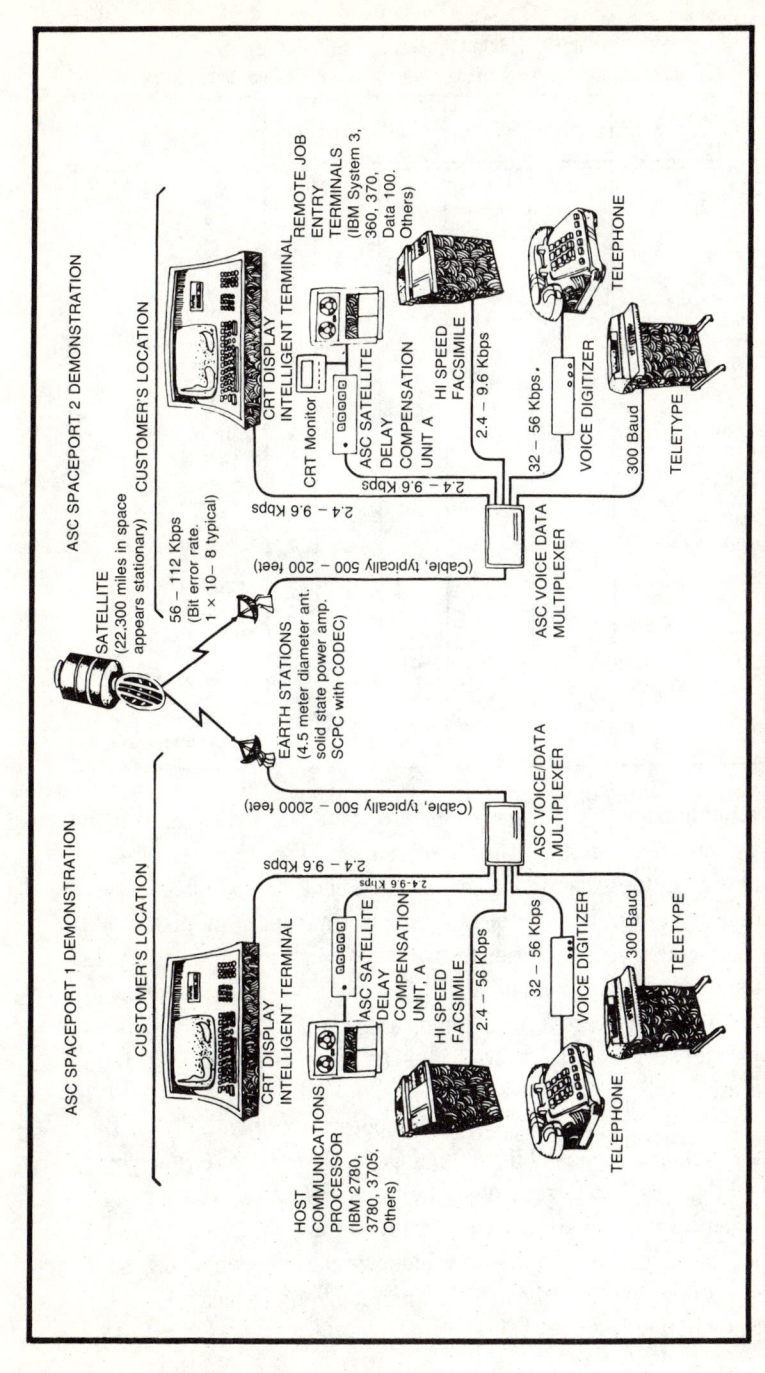

Fig. 3-6. ASC all-digit satellite service (Courtesy of Satellite Communications, Oct. 1977).

Table 3-1. Satellites Related to Telephone Communication (Courtesy of Satellite Communications, Oct. 1977).

Satellite System	Regional	Domestic	Fixed	Mobile	Exp.	Oper.
Comsat/AT&T		X	X		X	X 28,800 one-way telephony channels, or 1,000 mbs data
Comsat General/ Marisat	X	X		X		X 9 voice channels both ways 110 teleprinter channels-both ways.
Sirio (Italy)		X			X	X 12-100 kHz phone channels
Symphony (Germany/France)	X		X		X	X 1200 one-way telehpony.
						200+ phone circuits
USSR	X		X			X 12 video channels (one way)
Western Union		X	X			X 14,400 FDM voice channels (one way)

is orbiting over the Indian Ocean, providing service to more than forty countries. In 1976, the three Intelsat IV-A satellites already operating in the Atlantic Ocean area handled over 50 million calls. This new Intelsat IV-A satellite is able to transmit more than 6,000 simultaneous calls, in addition to two television transmissions. Table 3-1 summarizes the satellites related to the telephone communication. Since computers form an important part of today's communications media, let's take a brief look at computers and how they operate.

As the name implies, digital computers work with digits. But since a computer doesn't have ten fingers, there is no reason why it should have ten different digits. Computers work with just two different digits, 0 and 1, because they are easy to build this way. Since these digits have only two values, we call them binary digits, or bits. Putting several of these bits together gives us a binary number, such as 010 or 11001011.

The reason bits are used rather than the decimal digits 0 through 9 is that a 0 or a 1 is very easy to represent in an electrical circuit. Inside the computer, the 1 may be represented by a voltage on a wire, while the 0 is represented by either no voltage, or a very small voltage. Following the same idea, bits may be stored in a punched card as a hole for a 1, or no hole for a 0. In general, it's easy to store a 1 as the presence of something, while the 0 is the absence of it. This can apply to the voltage on a circuit, a hole in a card, a magnetic field in a piece of recording tape, or a light in a light bulb. Since there are only two possible digits (bits), there is little chance for error—the bit is either a 0 or a 1, with no digits between.

Digital data can be transmitted at many different speeds, and individual devices can frequently operate at two or more rates. These rates can be expressed as characters-per-second (CPS), words-per-minute (WPM) and baud or bits-per-second (the term baud is used interchangeably with bits-per-second).

The communications interface is the electrical connection between the computer and the communications equipment. The latter is generally a data set or modem, the device mentioned earlier that converts DC signals into AC signals suitable for transmission over telephone circuits, and then converts them back to DC signals for reception. The computer equipment is either an in-output port, or a terminal. It is fairly common for two pieces of computer equipment to be connected together, at which time they are called "data terminal equipment" and the "data communications equipment."

The heart of the computer is the CPU, Central Processing Unit, or *microprocessor*. This tiny, nifty device performs many functions, such as adding, subtracting, and comparing in response to instructions. In other words, it can do many things, but it must be told what to do by giving it an instruction. It is a digital device, so when it adds or subtracts, it does so in binary numbers, and the instruction given to it must be in the form of a binary number.

If the microprocessor has one thing going for it, it's speed. It can perform many operations in a very short time. If the microprocessor is to perform quickly, it must be fed the instructions just as fast. The micprocessor is connected to a memory where instructions and the necessary data are stored in the form of programs. Programming a computer means that you have to write, enter, and then store the program in the computer's memory.

Programming a computer and linking the memory to a telephone—your telephone—expands the limits of your telephone almost beyond belief.

Thus, the world will become smaller and smaller, our phone calls will be handled faster and faster, and hopefully, less expensively.

COMPUTERS AND THE TELEPHONE

As we learned earlier in this Chapter, digital communications are important to telephone technology. And since computers work with digits, their interfacing with telephones becomes an important facet of telephone communications.

Presently in England an experimental "Viewdata System" is connecting a telephone to a TV set which will link 1,000 homes and offices with the future. The occupants of the 700 homes and 300 offices will be part of a trial program by the post office (in England the post office handles the government owned telephone company). Through a slightly modified television set, the people will be able to find out about such matters as houses for sale, welfare benefits, train schedules, and stock market transactions. They can also reserve airline tickets, theatre and movie tickets, calculate income tax and mortgage payments, and leave messages on the screens for other users across the country.

Users will choose pages for their TV screens using a remote control device which resembles a calculator. The post office will not supply the information, but will leave that to such organizations as the *Times* newspapers, London Transport, British Airways, the Library, the government's central office of information, and Reuters and Associated Press. These firms will pay for the privilege of putting their information out and will in turn bill the home and business users. Users will be told how much they have spent after each session with the computer. They will also be paying standard phone rates for the call which links their set with the computer. An important part of this trial is to find out who is willing to pay for what information.

A potential future development for this type of system is a printer by which users could get printed copies of materials they want (Fig. 3-7). There are also plans to provide a system allowing

Fig. 3-7. Obtaining copies of requested material via telephone-computer linkage.

users to punch credit card numbers into the computer, thus not only locating goods and services, but also paying for them as well. The information services available could extend themselves to the following:

    Plays and movies from a video library
    Computer aided school instruction
    Cashless society transactions
    Computer tutor
    Adult evening courses on TV
    Correspondence school
    Dedicated newspaper
    Answering services
    Computer assisted meetings
    Household mail and messages

Fig. 3-8. Checking the menu at a restaurant via telephone-computer linkage.

Shopping transactions
Banking services
Grocery price list, information and ordering
Special sales information
Consumers advisory service
Daily calendar and reminder about appointments
Legal information
Weather bureau
Fares and ticket information
Library access
Restaurants
Mass mail and direct advertising mail

All this, and perhaps more, by simply pushing the buttons of your telephone (Fig. 3-8).

Figure 3-9 pictures VideoBrain's™ Family Computer VB-1000 Money Manager program cartridge. This program enables the user to keep track of tax-deductible expenses and balance checkbooks.

Apple II's next household appliance (Fig. 3-10) is a home servant to water your lawn, monitor the cooling and heating of your

Fig. 3-9. The Video Brain™ Family Computer Money Manager program. (Courtesy T.J. Ross and Associates, Inc.).

Fig. 3-10. Apple II, a future home servant.

house to save money, remind to make payments, and warn you when you are spending too much on entertainment.

The future of telephone—computer linkage—mindboggling!

FCC Rules and Regulations

It has always been possible to own your own phone and not lease it from the telephone company, but this has not been generally known. However, very few individuals have been able to do this because the phone company has always charged a fee for special "protection devices" that make owning your own phone uneconomical.

Now, however, as a result of an FCC (Federal Communications Commission) order, everyone will be able to buy and install their own phone without having to pay a special fee. All that will be necessary is a jack supplied by the telephone company to connect the equipment to the company network.

MANUFACTURER AND SUPPLIER REGISTRATION

The FCC's theory is that if equipment meets the registration requirements, it has little potential for harm to the telephone network, or for degradation of services to other users. However, the technical requirements contain no specifications in many important preformance areas. Such things as dial speed, percent break, tolerance and amplitude of DTMF tones (Dual Tone Multiple Frequency, or Touch-Tone^R), and even the fidelity of the transmitter and receiver are not specified.

FCC registration is like an Underwriter's label on an electrical appliance. The UL label on an electric drill does not testify to the quality of the drill. It only indicates that it probably won't catch on fire or electrocute the user. Likewise, FCC registration will not guarantee that a telephone is a quality product which will perform well for many years. It simply indicates the telephone probably will not cause disruption of service to other telephone users, will not endanger telephone company service people, and will not cause malfunction of the nationwide telephone switching network (Fig. 4-1).

Manufacturers and suppliers of telephone equipment must register new equipment with the FCC. Each application includes test procedures and results of performance data and technical material to assure compliance with FCC's technical standards. Public notice is made of the application to give any interested party the opportunity to review the specifications and, if necessary, to file comments.

Manufacturers, wholesalers, retailers, refurbishers, importers, even individual users, can apply for and obtain registration. If the FCC approves the application, it will furnish a registration number which the registrant must put on the device. When purchasing equipment from a manufacturer or supplier, always check to see if the equipment has the appropriate FCC approval. Buying non-approved equipment will get you and the supplier in trouble, and may damage the telephone company's network.

As part of the required label, the manufacturer or supplier must also furnish the *ringer equivalence* number, which is needed to determine if a device can be connected to a particular customer line. There are approximately fifty possible ringer variations presently in use throughout the U.S. These vary from telephone company to telephone company, and from private line to party line. If you connect an incompatible ringer to your phone line, you can cause all kinds of service disruptions for both yourself and your neighbors, possibly resulting in repair charges from the phone company.

Many manufacturers and suppliers are selling phones without ringers. With these many ringer variations, there is no point in predetermining ringer configuration for phones. Furthermore, most people buying a telephone are actually seeking the convenience of an extension phone and do not need the bells to ring in their home anyway.

The manufacturer or supplier must provide installation instructions and operation and repair procedures for the user. If registration of the equipment is revoked, the manufacturer or supplier must

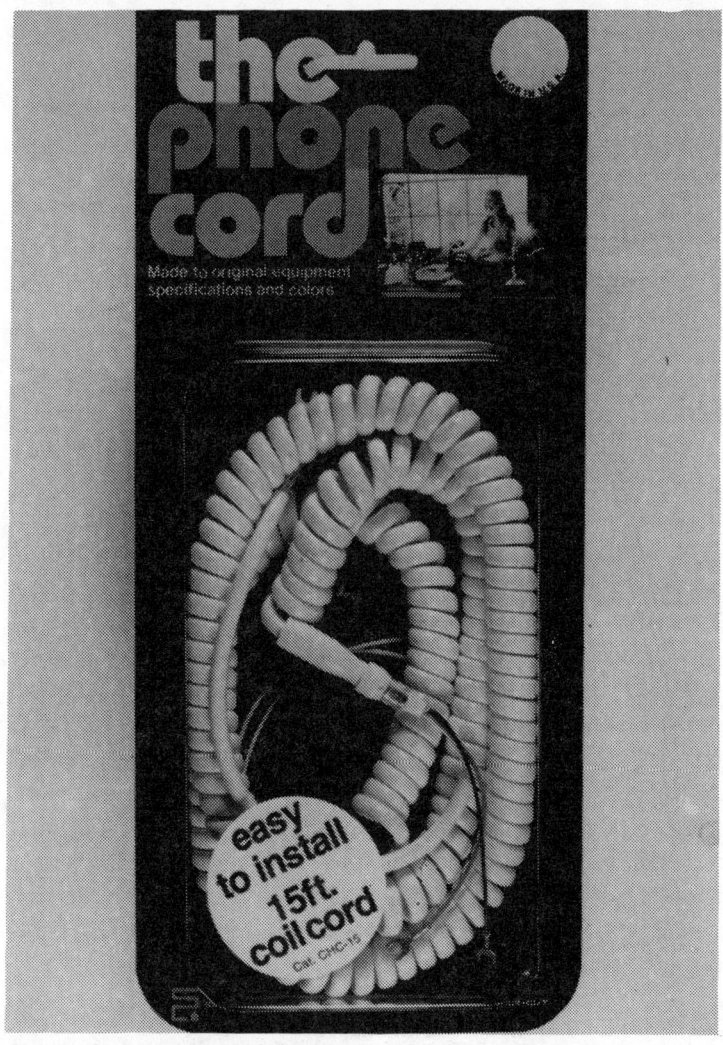

Fig. 4-1. Commercial packaging of phone equipment, in this case a fifteen foot line cord (courtesy of Smith-Gates Corp.).

take all reasonable steps to notify purchasers about the revocation and to instruct the purchaser to discontinue use of the equipment.

OWNER REQUIREMENTS

If the customer does not already have a telephone company jack, he must order one, and pay for its installation, depending upon the local tariff. If the customer already has a jack, he must still notify the telephone company of his intent to connect a registered device to the jack, and must identify the telephone line, furnish the device's registration number and ringer equivalence number, and furnish the vendor-supplied USOC code of the jack and the wiring arrangement required by his device. The customer must follow the instructions of the manufacturer or supplier, and comply with FCC tariffs regarding the use of registered equipment. Figure 4-2 illustrates some of the modular equipment available to the customer.

If the customer has trouble with this phone, he must disconnect the registered equipment. If the equipment is determined to be malfunctioning, it may not be used again until the problem has been corrected. Repairs may be performed by the customer if the instruction manual provides details on repair procedures. All other repairs must be made by the manufacturer, assembler, or other authorized agent. Even though the equipment is registered, the phone company can prevent you from connecting it to their network if they can show that the use of the equipment interferes with the service to others, or is in violation of their tariffs.

You may connect registered equipment, or connect unregistered equipment through *registered protective circuitry*. You may connect customer-owned equipment to a private line, but you may not connect customer-owned equipment to party lines or coin telephone lines. Phones on party lines are specially wired and use special apparatus to allow the central office to identify which party on the line is making a call. Use of standard phones can cause billing to the wrong party and failure to complete calls. Furthermore, special ringing schemes are used on party lines.

TELEPHONE COMPANY REQUIREMENTS

All terminal access points must have a jack at the end of the wire, and according to FCC rules, the telephone company still does the wiring and furnishes the jack. The jack and plug is the interface between the phone company and the customer. The telephone company may discontinue service if it is determined that the equipment the customer has provided causes harm to the phone company's network. Figure 4-3 illustrates equipment available from GTE Electric for use with customer-owned phones.

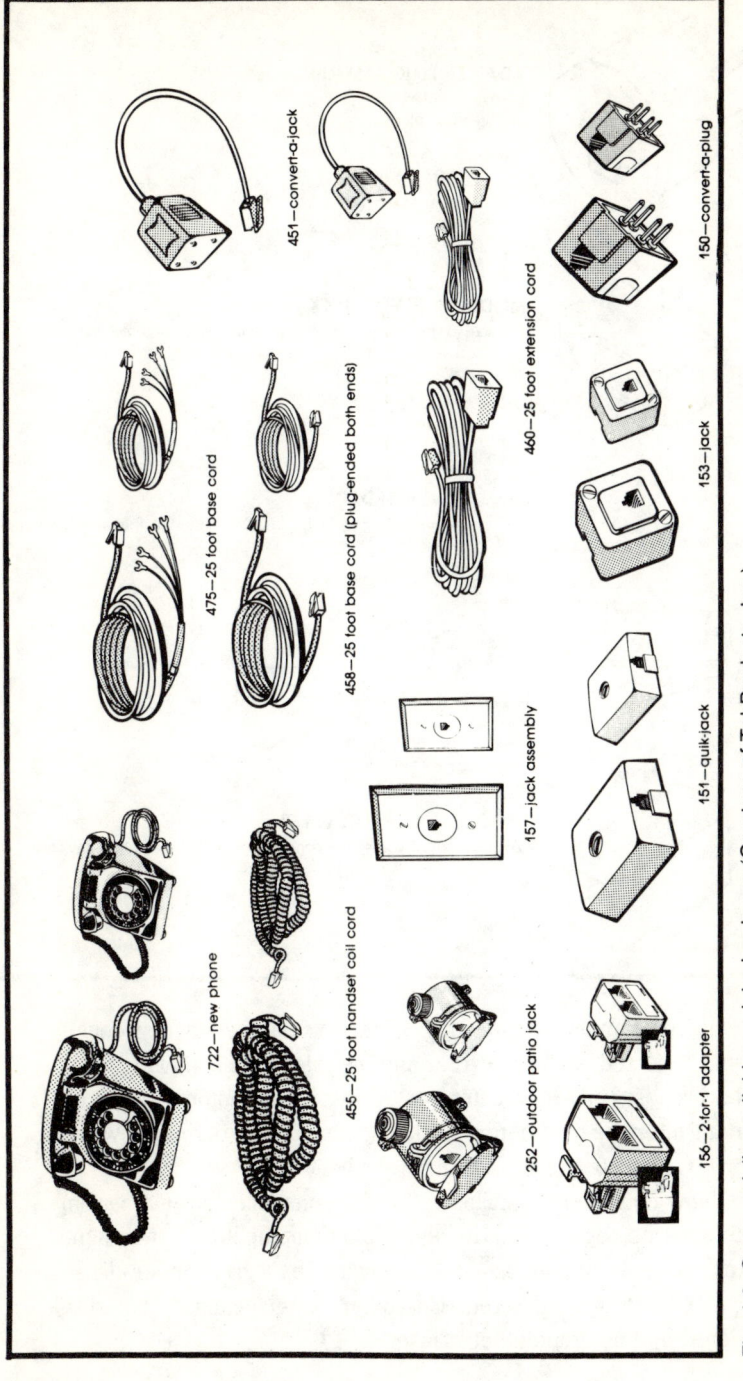

Fig. 4-2. Commercially available modular hardware (Courtesy of Tel Products, Inc.).

65

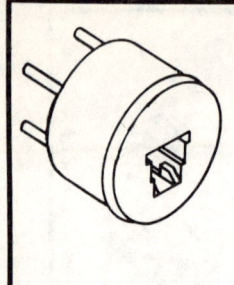

ADAPTER PLUG ASSEMBLY
Permits use of modular plug-ended cords when older four-point jacks are already in place.

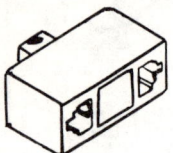

MODULAR DUPLEX JACK
Allows two telephones to be plugged into a single modular jack.

LINE CORDS
Modular plug on both ends. Transparent aluminum color. Available in 7', 14', and 25' lengths.

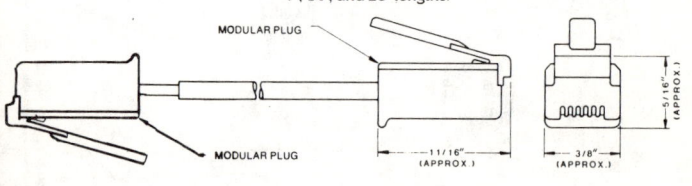

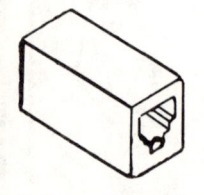

MODULAR CORD COUPLER
Allows modular cords to be connected together to lengthen line cord.

Records must be kept of registered customer-owned equipment connected to the network, and the telephone company must, upon request, provide customers with technical information concerning interface parameters, including the number of ringers which may be connected to a particular telephone line.

New equipment installed by phone companies after a certain date must be registered. However, old equipment that the phone companies have already installed is covered by a grandfather clause and need not be registered. Customer-owned equipment is also covered by this grandfather clause.

FLUSH JACK

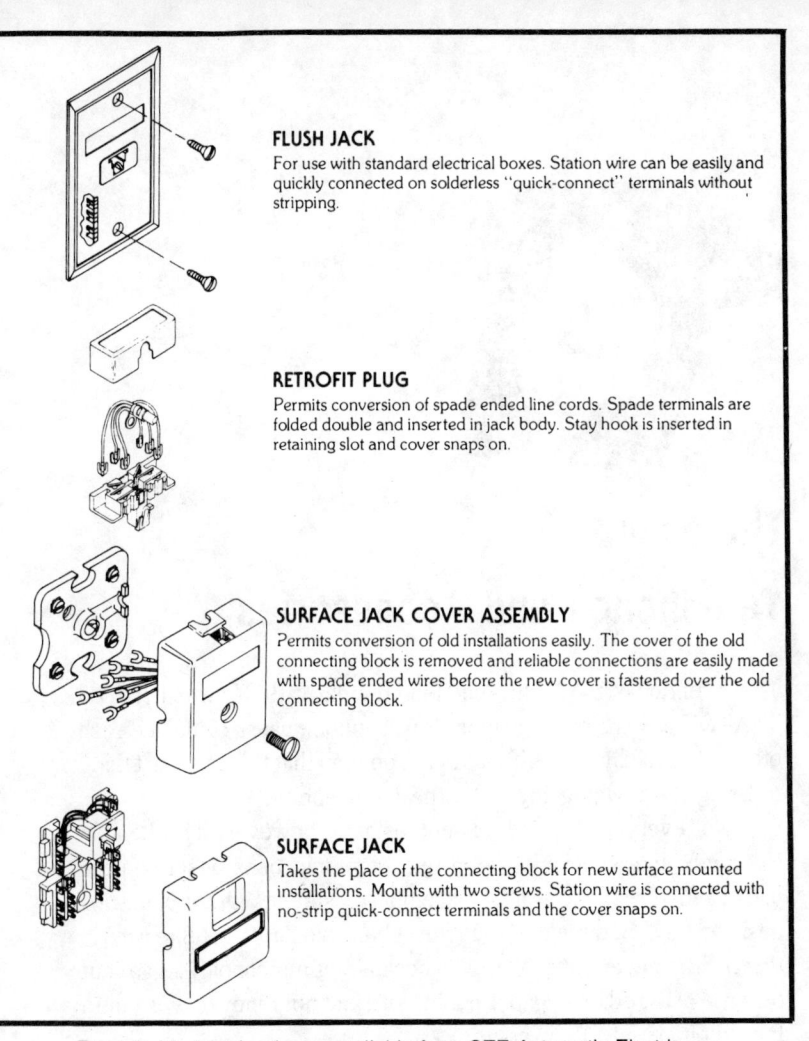

For use with standard electrical boxes. Station wire can be easily and quickly connected on solderless "quick-connect" terminals without stripping.

RETROFIT PLUG

Permits conversion of spade ended line cords. Spade terminals are folded double and inserted in jack body. Stay hook is inserted in retaining slot and cover snaps on.

SURFACE JACK COVER ASSEMBLY

Permits conversion of old installations easily. The cover of the old connecting block is removed and reliable connections are easily made with spade ended wires before the new cover is fastened over the old connecting block.

SURFACE JACK

Takes the place of the connecting block for new surface mounted installations. Mounts with two screws. Station wire is connected with no-strip quick-connect terminals and the cover snaps on.

Fig. 4-3. Modular hardware available from GTE Automatic Electric.

The FCC has adopted a group of standard plugs and jacks to be used. Telephone instruments and ancillary devices are to be connected through six or eight conductor modular plugs and jacks, or through fifty point microribbon connectors. If the telephone company makes any changes in facilities, equipment, operations, or procedures which may affect the operation of customer-provided equipment, it must notify the customer in writing.

The New Telephones and Accessories

The invention of the Picturephone in the early 1970's was heralded by AT&T as a major innovation. Finally phone users could see each other while talking. AT&T was of the opinion that the next best thing to being there was going to be the Picturephone call.

Not everyone shared this enthusiasm, however. The London *Economist* in a special issue on telecommunications described the use of a Picturephone as a "social embarrassment." Other countries besides the U.S. developed a Picturephone similar to the one used in the U.S. It never caught on, not because of a technology gap, but because of a credibility gap. People were just not ready to welcome a Picturephone into their homes.

THE PICTUREPHONE

Experts believe that the Picturephone (Fig. 5-1) will catch on as an economically viable consumer item during the early 1980's. The design is one of the most impressive pieces of modern electronic engineering.

The unit consists of three components: a conventional Touch Tone[R] telephone; a picturephone set with its screen, camera and loudspeaker; and a control unit which contains a microphone and which permits you to adjust your picture and speaker volume. There

Fig. 5-1. AT&T's Picturephone.

is also a separate control unit containing the power supply and line interfacing electronics. The system specifications are:

Bandwidth .... 1 MHz
Screen size .... 5½ inches × 5 inches
Frames per second 30
Number of lines per frame 250
Normal viewing distance .... 36 inches
Normal area of view 17½ inches × 16 inches to 28½ inches × 26 inches

The control of the network and the signals used for its control and for dialing are basically the same as for the conventional phone described earlier. What is added are wires to carry the picture in parallel with the existing local loops from the central office to your premises; new switching facilities operated in parallel with the telephone switching, but under the same control; and trunks capable of carrying the Picturephone signals as well as speech (Fig. 5-2).

All Picturephone transmissions between central offices will be done in digital format. Once digitally encoded, the signals will remain in that form until they reach their destination office, or the closest office to it that can decode them by means of a modem into analog form (Fig. 5-3).

Presently, Picturephone is likely to remain too expensive for your home. Picturephone booths in large cities may become popular first, as an intermediate solution, for talking to faraway friends or relatives. Video conferences in industry and for private use are already in existence, linking New York, Washington, Chicago and San Francisco. Called PMS, Picturephone Meeting Service, the system enables a group of people to conduct meetings without the need to travel.

PMS centers are specially designed rooms that can accommodate six active participants, and a number of other participants who sit off-camera during a meeting. These systems are designed for maximum comfort and efficiency and include a range of graphic capabilities that make possible easel presentations and the use of slides, transparencies, video tape, and even close-ups of sketches and handwritten notes (Fig. 5-1). Some the the features of the service include cameras that automatically switch to the person

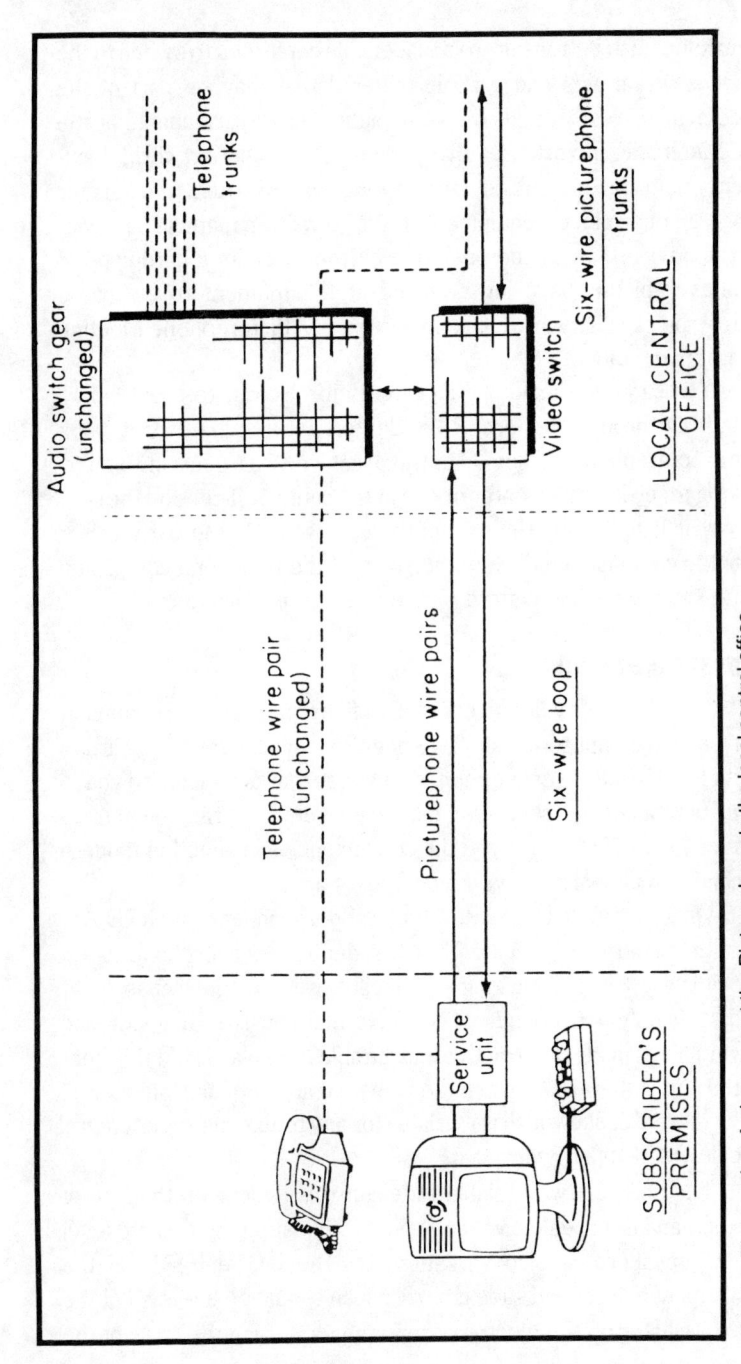

Fig. 5-2. Three pairs of wire connect the Picturephone to the local central office.

speaking; push button controls that enable conferees to switch to the various visual aids and graphic material that may be part of the meeting, or to select an overview picture of all participants at the meeting table; a work area that permits showing of artwork, documents, and transparencies with a tripod camera with zoom lens for close-up of products, equipment, or large written material; an electronic device that, at the push of a button, permits a hard copy of whatever picture is being received at the moment to be made; videotape capability in which any portion of a Picturephone Meeting can be videotaped.

You can even make arrangements for a coast-to-coast family gathering through the local PMS Center. It's not like having your own Picturephone at home, but at least you can see who you're talking to, be it at the Center or one of the central offices and the cost to you is usually only the cost of the call itself. By the early 1980s you'll have your own. That's the soonest the telephone companies think you'll be able to afford a Picturephone in your home.

THE SPEAKER PHONE

All the new developments such as fiber optics, digital communications and computerization have been responsible for the availability of hands-free "speaker phones," phones that don't require you to hang onto a headset while you talk. These devices, ranging in price from $150 to $250, offer you the convenience and speed of modern technology unheard of even a few years ago.

One such unit is the Residential Speakerphone by GTE Automatic Electric (Fig. 5-4). This Residential Speakerphone is approximately 8¾ inches long × 6 inches wide × 2⅝ inches high, weighing only 1.6 pounds. The enclosure consists of upper and lower halves made of attractive and durable ABS plastic. At the front of the unit is the *on-off* switch (hookswitch) and *mute* button, as well as an unused space which is available for any future customer control should the demand warrant it.

The unit has two variable slide controls. One is on the face of the unit and is a speaker volume control, adjusting the receive level of the speakerphone conversation. On the left rear side of the instrument is a second slide control which regulates the level of the tone oscillator on the electronic tone ringer. Lifting the shaft of the

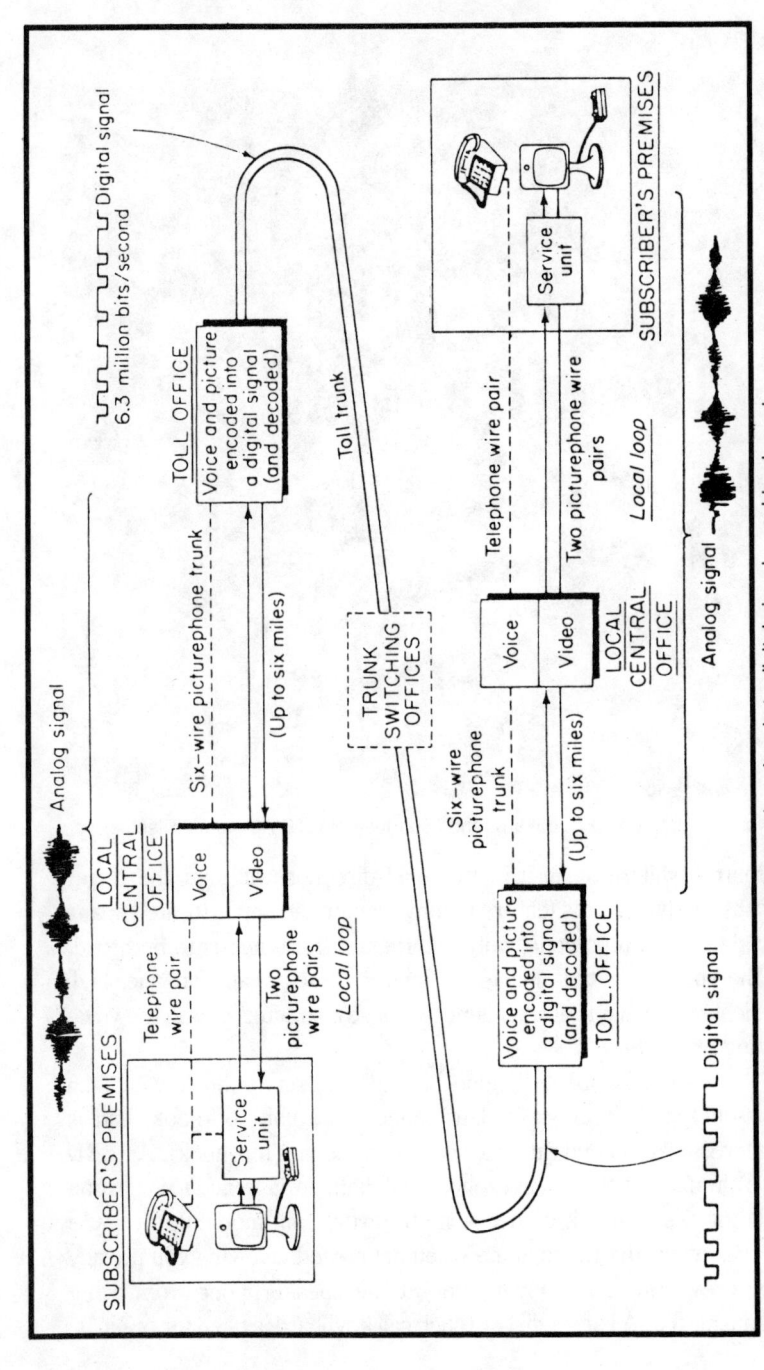

Fig. 5-3. The picturephone necessitates translating analog signals to digital signals, and back again.

Fig. 5-4. Hands-free Speakerphone (Courtesy of GTE Automatic Electric).

control while moving it in the *quiet* direction will shut the ringer completely off. This feature is convenient if the unit is located near a conventional telephone; only one ringing signal needs to be heard. The speaker in this unit is a 2 inch × 3 inch oval, dynamic, high fidelity unit which has excellent response characteristics in the voice range (Fig. 5-5).

Since this speakerphone has no handset or hookswitch, an on-off switch is depressed to connect the unit (off-hook) and is depressed a second time to disconnect the unit (on-hook). An LED on the face of the speakerphone will light to give an *on* indication during the off-hook mode. A mute switch will momentarily mute (deactivate) the microphone when depressed, allowing you privacy by preventing any conversation at the speakerphone from being transmitted. A rotary dial or touch call keyboard is used for originating your call. You simply depress the on-off switch and dial, or push, the number you want when you hear the dial tone. When your party answers, you may adjust the speech level with the volume control. Go about your business and converse, from any distance up to twenty feet, as you would with any standard phone. When you're finished talking, walk to the unit and depress the on-off switch. That's it, simple and no fuss of holding a handset in your hand.

THE ELECTRONIC TELEPHONE

Another new device is the *electronic* phone (Fig. 5-6). The Adonis-G by Allied Telecommunication Equipment is one version of the electronic phone. It is a single-piece, lightweight unit that allows hands-free operation. The miniaturized integrated circuits have eliminated over 120 parts found in standard phones, and it has a memory circuit which stores the last number called.

In this type of unit, the conventional electromechanical components such as the bell ringer, the transformer-coupled speech network, and the tone oscillator dial have been replaced by integrated-

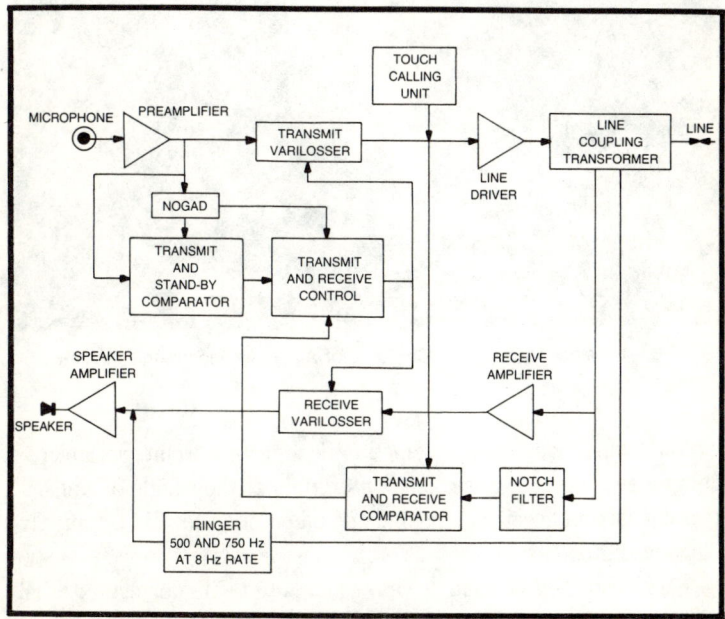

Fig. 5-5. Circuit diagram of the Speakerphone (Courtesy of GTE Automatic Electric).

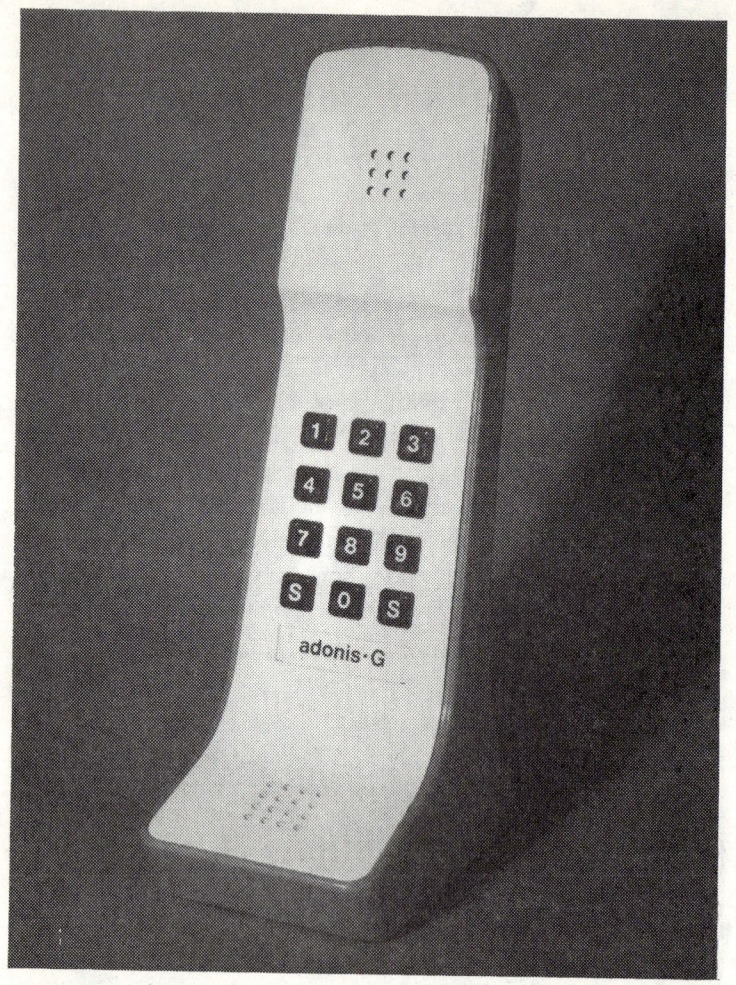

Fig. 5-6. The electronic phone (Courtesy of Allied Telecommunication Equipment).

circuit "chips," one for each function. Integrated circuits and micro-components make it possible to interconnect thousands of transistors, diodes and resistors on a single piece of silicon chip, typically 0.15 inches square.

The three chips used are interconnected with certain external components to make up the basic telephone set. Because of the chips and the micro-components, the shape of the phone can be redesigned in a variety of ways, limited only by the imagination of the designer. Many features that in the past could only have been obtained with expensive additional equipment can now be an integral part of the electronic phone.

The Wonder Phone (Fig. 5-7) by ATE, (Allied Telecommunication Equipment) functions as a calculator as well as a phone. To operate the unit as a phone you lift up the handset, wait for the dial tone, then depress numerals for the number you want to call. When you caller's phone number is busy, you don't need to dial his or her number over and over again: you depress the # key, leave the handset off the hook, and the phone will automatically dial the number. In the case that the party you've called leaves a message to call back in, say, five minutes, you can do that by depressing the T button. You can also program the phone to remember the sixteen phone numbers you most frequently dial.

The Wonder Phone functions as a calculator as well as a telephone. The keyboard consists of:

0 – 9	Numeral key
+ – × ÷ = . +/– %	Calculator function key
E	Entry clear key
C	All clear key
#	Reset call key
*	Short dial outgoing key
H	Time setting clear key
T	Time setting key
W	Changeover key

You can calculate eight figures either on line busy, or as an ordinary calculator:

    0–9, the numeral keys: use these keys for calculation.

    ., decimal point key: use for entering decimals into calculator, for example 12.3 is 1 2 . 3, and 0.1 is . 1.

    +, –, ×, 6, =, basic calculator keys: use according to mathematical formula in addition, subtraction, multiplication and division, using the = key to find the calculated result.

    C, clear key: Depressing this key will clear all figures.

    E, entry clear key; used immediately after a wrong number is dialed, clearing the latest number from the memory.

Fig. 5-7. The Wonder Phone by Allied Telecommunication Equipment is a calculator/time/memory phone.

%, percent: used for ordinary percent calculations, 200 × 3% = .

×/−, sign clear key: used to discount calculations, 5% discount from 100 = 1 0 0 × 5 % +/− = .

There are a number of telephones of this type on the market. Remco International, for example, has a similar phone as described above, without the calculating feature. Their Super Phone allows for hands-free dialing, automatic repeat dialing in case you encounter a busy signal, and a computerized memory for ten phone numbers.

Then there is a Data Phone (Fig. 5-8), from Data Phone Corp., which features an automatic display of the hours, minutes and seconds for the duration of seven seconds, and the month and day for three seconds. This time display will disappear from the LED display when the unit is used for other purposes, but it will return. This telephone also has an alarm clock, with an extra "snooze time." You can also accurately record time spent on phone conversations, and there is an automatic recall of the last number called.

Fig. 5-8. A Data Phone, a compact, multi-purpose, easily installed memory bank telephone and calculator by Data Phone Corporation.

Electro-Atomic Products Corp. manufactures the Compuphone which offers a memory storage feature of 100, twelve digit telephone numbers, each of which can be recalled by a two number digit code entered through the keyboard. This phone also offers automatic and manual dialing, visual readout, battery backup, and stopwatch and a time-zone display, giving the local time plus two additional time zones.

Phone Devices Corp. supplies the Model 700 Figure-Phone, a first generation calculator/clock/calendar phone combining the phone with a handheld calculator, a digital alarm clock and a daily calendar.

Consumers are not specifically aware of how they are billed for phone service. You may waste a significant amount of dollars in the average billing period because you lack proper time period utilization. You are charged for fixed intervals of time, and you are billed in advance for each time period during the progression of your call.

Fig. 5-9. The Computer • Phone 370 allows you to keep accurate account of the cost of your phone calls (Courtesy of Utility Verification Corporation).

When you terminate a call twenty seconds into a new time period, you are billed for the entire period. On an average, you do not use more than 50% of the last charged billing interval.

The telephone companies' tariffs are complicated and based on state and federal tariffs, mileage, time of day, day of week, and other factors. As an average phone user, you may have a poor concept of the relative cost of your phone calls.

The Computer • Phone 370 (Fig. 5-9) is a truly unique device in its ability to display the true cost of a call before and during your calls. The device uses the latest microprocessor technology and solid-state construction to combine aesthetic appeal with the convenience of pushbutton entry for pulse dialing of calls, with remarkable functional capability to serve as a cost-saving product.

When not in use as a telephone, the bottom display serves as a clock using European numbering, that is, 11:00 PM = 2300. To set the clock, depress letter C and number 1, enter the time as hours, minutes and seconds (3:30 PM is entered as 15 30 00), and depress the key to complete the entry. To set the calendar, depress letter C and number 2, enter date as month, day, year (December 1, 1977 would be entered as 12 01 77) and depress # key to complete the entry. To enter your phone number, depress letter C and number 3, enter your area code and phone number, and depress # key to complete the entry.

To make a phone call, you dial the number you wish to call as you would on any Touch-Tone[R] phone. Then depress the * (asterisk) key and the computer will do the rest. It will display the initial billed cost period and the period of time for which you will be billed. It will display secondary (next period) cost and the time period for which you will be billed. It will maintain the accumulated billed cost and billed time. When there are fifteen seconds left in the current billed time period, the displays flash a warning to you that soon you will incur the charge for the next full period. Upon connection, the bottom displays convert to a timer.

You can even determine the cost of a call without making the actual call by depressing button D and dialing the number. Depress * key, and the displays will show the initial period billed cost and the secondary cost and time period for which you will be billed. You can further determine the accumulated cost and time by depressing letter C and number 4. The accumulated billed cost and billed time will show on the upper displays. Depress the # key to release the display.

The Computer • Phone 370 will redial the last number, and will repetitively redial a busy number. When the connection is completed, a bell will sound to let you know that the number is not busy any longer.

The TeleCoster I (Fig. 5-10) is a simpler, non-automatic version of the Computer • Phone. When you purchase this unit, a Call Cost Directory is supplied which lists day, evening, nighttime, and rate periods from your telephone number to anywhere in the continental U.S.

To use the TeleCoster I, you look up the area code and exchange which you are calling in the Call Cost Directory, and enter the initial and secondary time and cost data into the machine, and press the *start* button when the call begins. The unit displays the costs and time spent on the call, and will flash a warning light fifteen seconds before incurring the additional cost period. You can see what the next period cost would be anytime during the call.

Fig. 5-10. The Telecaster I, a simpler version of the Computer • Phone (Courtesy of Utility Verification Corporation).

Owners of Computer • Phone and TeleCoster I will be kept advised as to phone company rate changes which may effect either of the products.

Another digital timekeeping device is manufactured by Cervco, called the Digi-Toll. The unit is not connected to the telephone system, but its special circuit continuously displays the same time-of-day and elapsed-time data as used by the phone company to calculate toll charges.

The Digi-Toll is operated from a standard 115V AC outlet and functions as a twelve hour time-of-day clock and twenty-four minute call timer.

An LED display provides easy readability. Important timing and discount-data tables are prominently located above the LED display for easy reference. These tables are supplied by the manufacturer.

CORDLESS TELEPHONES

Portable, cordless telephones (Fig. 5-11) available from such suppliers as Fracom Enterprises, Allied Telecommunications Equipment, Tel-Tech Inc. Com/Link International, Gutzmer International and Davis-Denver Co. operate on a combination of telephone and radio principles. A portable phone is powered by re-

Fig. 5-11. A cordless telephone with base unit (Courtesy of Fracom Enterprises).

chargeable batteries and transmits a private two-way conversation by FM signal to a base unit. This base unit plugs into a phone jack and the usual 115V electric wall outlet. The average operating distance of a cordless phone is 300 feet from the base unit.

The following, specifications of a cordless telephone from Fracom Enterprises are typical. The phone unit contains rechargeable Nicad batteries (7.5V DC) with a rear mounted on-off switch, and operates without recharge. Standby is up to twenty hours; talk is up to eight hours. There is an audible tone ringing signal. The specifications are:

Transmitter section:
    27 MHz frequency
    100 milliwatts maximum power
    Narrow band FM
    Crystal controlled

Receiver section:
    1.7 MHz frequency
    Single conversion superhet.
    FM detector

The base unit is 13½ inches × 8¼ inches × 3¾ inches, with 115V AC, 60 Hz or 230V AC, 50 Hz, with a six-foot, three-wire, neutral ground line cord with a four-prong, five-foot standard telephone cord, and a line-matching hybrid/line-seizing relay. The specifications are:

Transmitter section:
    1.7 MHz frequency
    Carrier current
    Narrow band FM

Receiver section:
    27 MHz frequency
    Single conversion superheterodyne
    FM detector

Most cordless telephones are designed to operate only on single line (one telephone number) service. However, with some modifications by the manufacturer, connection to a two-line service is possible.

Before connecting any cordless telephone, you must give notice to the phone company as to the phone number to which the cordless phone will be connected, the FCC registration number of the cordless phone, and the ringer equivalent number. You must also give disconnect notice to the phone company when the cordless phone is removed from service.

Should the cordless phone cause harm to the telephone network, the telephone company shall, where practicable, notify you that temporary discontinuance of service may be required. However, where prior notice is not practical, the phone company may temporarily disconnect service immediately if such action is reasonable under the circumstances. In case of temporary discontinuance, the phone company shall promptly notify you of such temporary discontinuance, afford you the opportunity to correct the situation which has given rise to the temporary discontinuance, and inform you of your right to bring a complaint to the commission pursuant to the procedure set out in subpart E of Part 68 of the Telephone Act.

The telephone company may make changes in its communication facilities as per Part 68.110, in equipment, operations or procedures, where such action is reasonably required in the operation of its business and is not inconsistent with the rules and regulations of Part 68. If such changes can be reasonably expected to render your cordless phone incompatible with the phone company's communica-

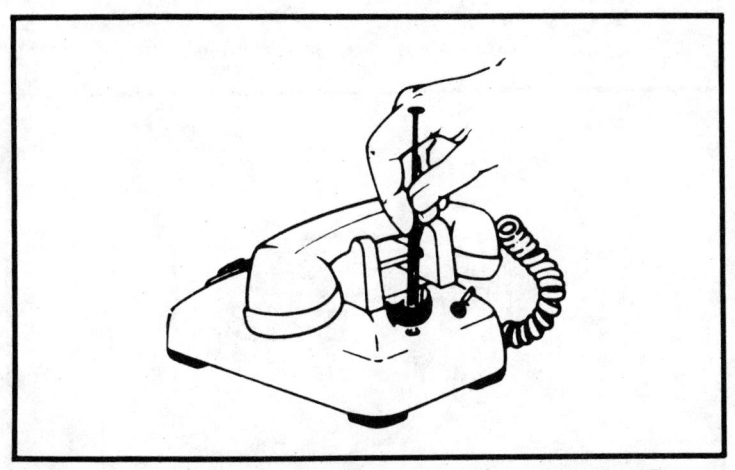

Fig. 5-12. The antenna inserted on back of portable unit (Courtesy of Fracom Enterprises).

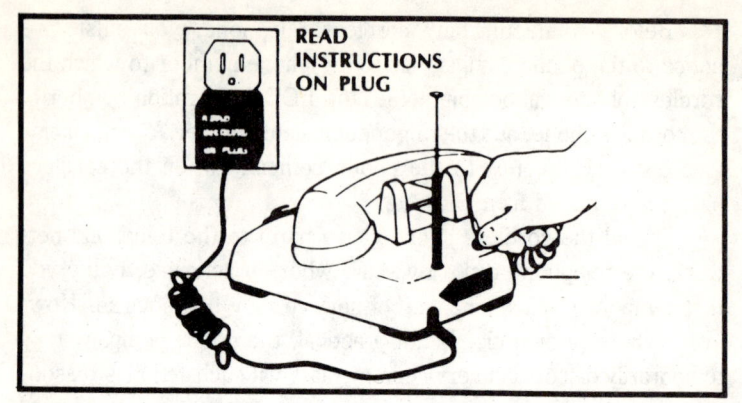

Fig. 5-13. Overnight charging of a cordless phone (Courtesy of Fracom Enterprises).

tion facilities, or require modification or alteration of the cordless phone, or otherwise materially affect its use or performance, you shall be given adequate notice in writing to allow you an opportunity to maintain uninterrupted service.

Installation of your cordless phone is very simple. Two antennas are needed to operate your phone, one at the base unit and one at the portable unit. Follow the manufacturer's instructions about installing these antennas (Fig. 5-12). The antenna is inserted in a hole on the back of the portable unit, turning clockwise until tight. Use fingertip pressure only. Do not use pliers or wrenches. Portable units using rechargeable Nicad batteries allow for overnight charging (Fig. 5-13). The on-off switch must be turned *off*. One overnight

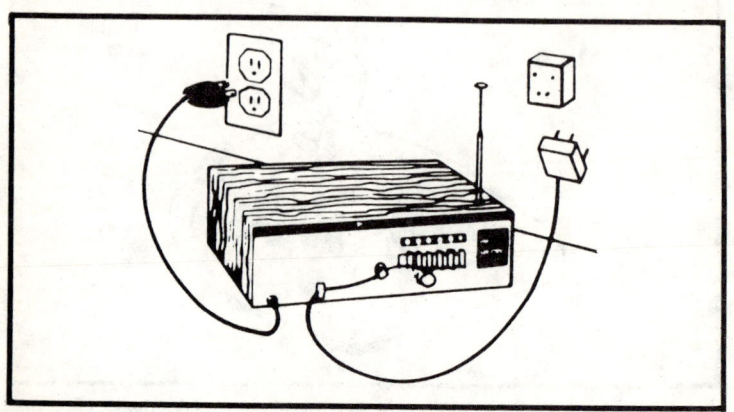

Fig. 5-14. Installing the cordless phone base unit.

charging will give a full day's normal use before recharging is necessary.

Connections are easily made. The cordless telephone, having no wires or other attachments besides the antenna, needs no installation whatsoever. The base unit is plugged into a standard 115V outlet, and the telephone extension is connected to a standard four-hole telephone extension jack (Fig. 5-14).

Since the cordless phone uses the airwaves for its communication link, interference from other sources is possible. This can be minimized by slightly lowering the antenna on the base unit. Also, the cordless phone signal can be directional, as it is in a portable radio. Merely turn the phone slightly to get maximum clarity. Figure 5-15 shows a portable version of a cordless phone.

Fig. 5-15. A portable cordless phone (Courtesy of Fracom Enterprises).

Allied Telecommunication Equipment manufactures a cordless telephone that incorporates a memory system (Fig. 5-16). In case you dial a busy number, press a memory key and go on about your business. The phone does the rest and it lets you know when connection has been established. You can also have the set memorize a specific number.

Royce Electronics supplies a cordless telephone which can answer any incoming calls with cord-free operation. Their cordless phones are perhaps the smallest ones on the market, measuring only 6 inches × 2½ inches × 1 inch.

MISCELLANEOUS ACCESSORIES

Accessories making your electronic phone dialing even easier consist of such items as a Call Diverter, Electronic Dialer, Pocket Yellow Pages, Fone Silencer, and Extension Bells.

Soft-Touch[R]

Soft-Touch (Fig. 5-17) instantly converts your standard dial phone into a modern Touch-Tone[R] telephone. The unit is actually a dial and microphone replacement in a standard phone mouthpiece. You simply unscrew your present telephone mouthpiece and screw on the Soft-Touch[R]. If the alignment of the numbers is not convenient, you can rotate the outer case until the numbers are in the right position.

Manufactured by Telephone Electronics Corp./Buscom Systems, the Soft-Touch[R] uses a calculator type of integrated circuit with over 4,000 transistors. It is crystal controlled with a tone accuracy of ±0.25%.

Call Diverter

Another device, made by Telephone Electronics Corp/Buscom Systems, is a call diverter (Fig. 5-18). It's not an answering machine (which is described in a later chapter), but answers your incoming calls by electronically diverting them to any phone number you select. These devices are ideal for persons who expect important calls and want to answer the telephone themselves, but will not be at home.

Fig. 5-16. Cordless telephone that incorporates a memory system (Courtesy of Allied Telecommunication Equipment).

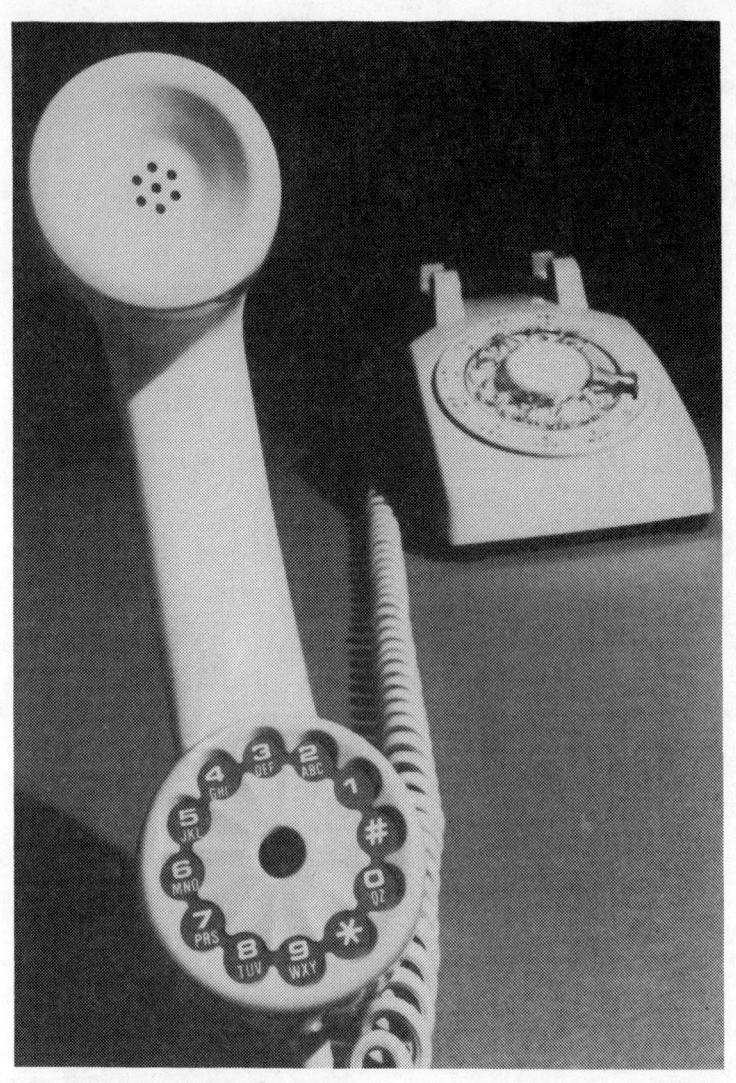

Fig. 5-17. The Soft-Touch^R converts your dial phone to a Touch-Tone^R one (Courtesy of Telephone Electronics Corp/Buscom Systems).

The installation of the call diverter requires close cooperation with your local central office. This unit incorporates *dial delay*, an adjustment of dial tone speed return that is available from the central office, and amounts to a time of zero to ten seconds, *speed*, an adjustment of ten or twenty pulses-per-second to match the speed

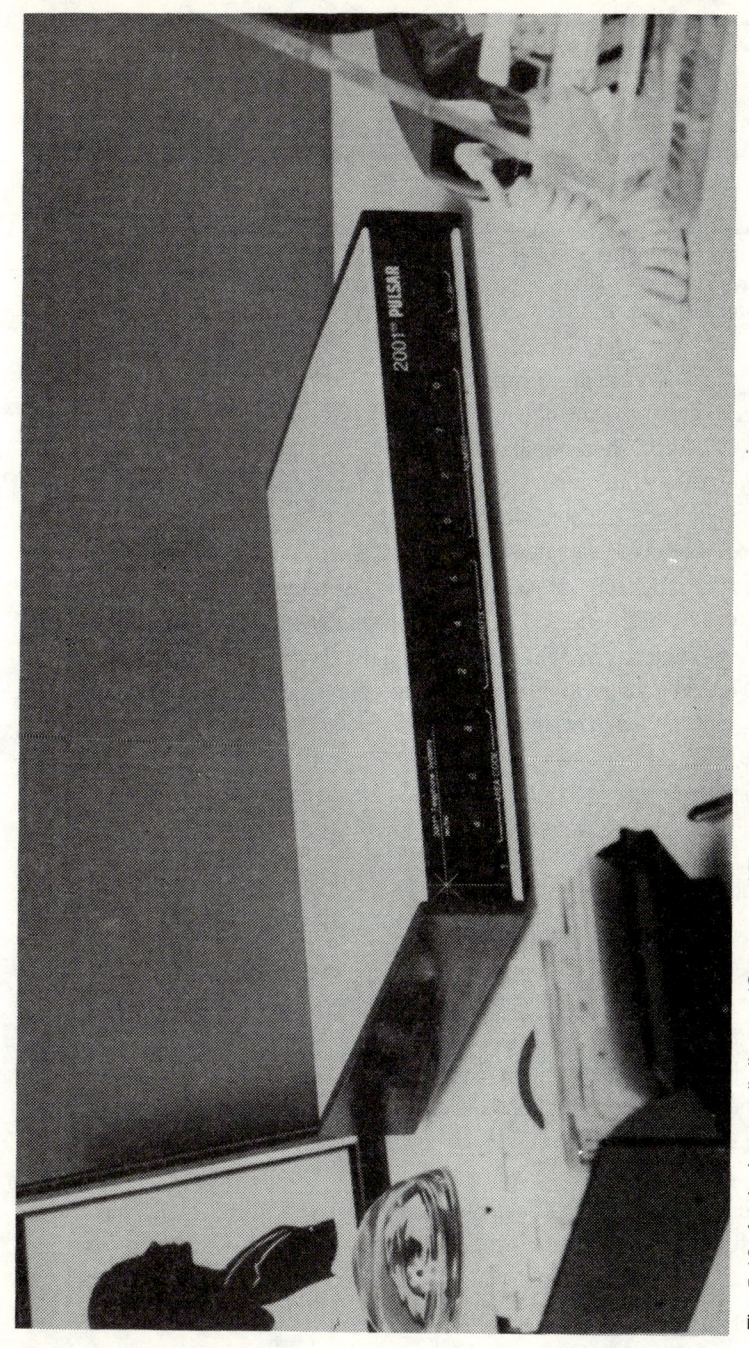

Fig. 5-18. A telephone call diverter (Courtesy of Telephone Electronics Corp./Buscom Systems).

91

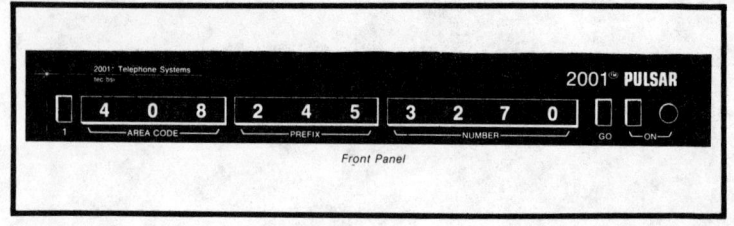

Fig. 5-19. Front panel of a telephone call diverter.

available from the central office, and *patch delay*, an adjustment for speed of central office switching, from zero to twenty seconds.

To operate this device, you need two telephone lines, or numbers. The incoming line into your call diverter is connected to your first, or published telephone number. The outgoing line from your diverter is connected to your second number.

Operation is quite simple. For areas requiring dialing of "1" for long distance calls, depress the "1" switch (Fig. 5-19). Enter if required the area code of the number you are diverting to; otherwise leave the area code at 000. Enter the seven digit number, and press the *on* swtich. The light is now on and the Pulsar is ready to instantly divert all your incoming calls out to your selected new number.

Ford Industries manufactures a similar device, except that they have added an answering feature to the diverter. The tape deck adapter gives your caller a recorded message, telling him or her to stay on the line while the call is transferred to another number.

Automatic Dialer

There was a time when making a phone call meant looking up the number, jotting it down, concentrating on the telephone dial, and carefully dialing each digit to avoid a wrong number. No more. Automatic Telephone Dialers make that a thing of the past. The dialers on the market store from sixteen to thirty-two of your most frequently called numbers and dial them automatically. At the touch of a button, you instantly reach any direct dial number in the world.

Ford Industries supplies a number of Code-A-Phone [R] Electronic Dialers. With Electronic Dialer # I you manually dial any of sixteen numbers on the programming panel, press a button to store it, and the number is permanently held in the memory until you wish to change it.

Electronic Dialer II has a built-in speaker that lets you hear the number being dialed and the phone ringing. If it's busy, or there's no answer, press another button to cancel the call.

Electronic Dialer III also has a built-in speaker, plus a LED digital readout of numbers during programming and calling, offering you a visual display of these numbers, just to make sure you programmed the right digits.

A more advanced automatic dialer is offered by Microelectronic Communications Corp (Fig. 5-20). Their unit offers thirty-two memory locations, automatic redialing of the last number, conversion of a rotary dial phone into a touch-type and a built-in calculator.

Micro-Dialer installation is simple. Whatever your phone jack configuration, your phone hooks up to the dialer and the dialer to the telephone network (Fig. 5-21). With the new modular wall jack system, Micro-Dialer plugs into the wall jack and your existing phone plugs into the back of the Dialer. When a four pin jack is

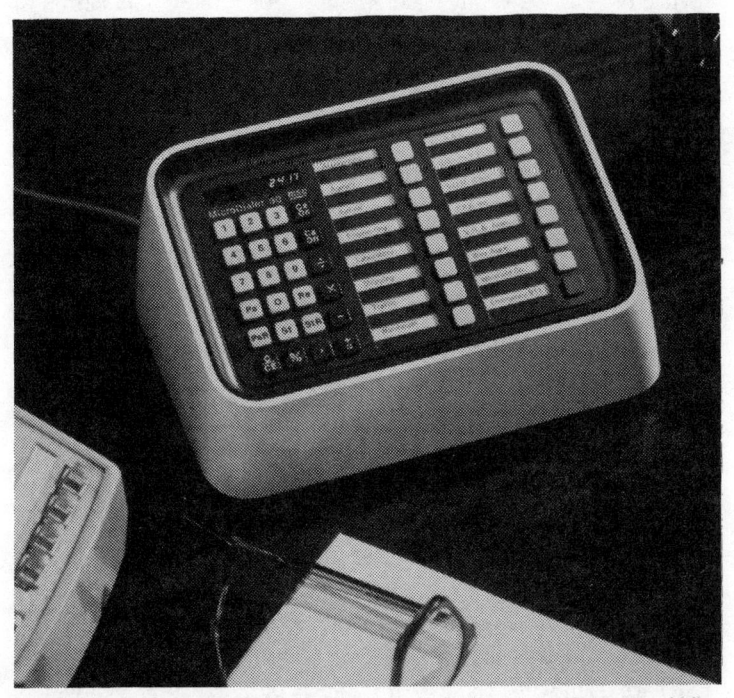

Fig. 5-20. Automatic dialer with thirty-two memory locations (Courtesy of Microelectronic Communications Corp.)

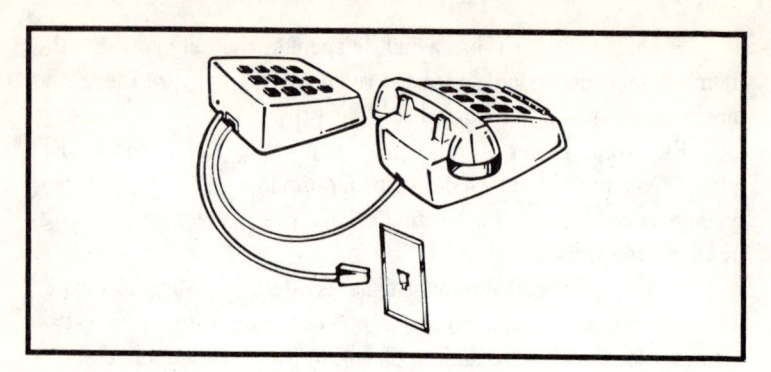

Fig. 5-21. Micro-Dialer plugs into your existing wall jack.

installed at your existing phone, use an adaptor cable to connect your phone to the Dialer, and an adaptor plug to connect the Dialer to the telephone network (Fig. 5-22).

If you do not wish to use the rotary dial or touch-tone pad on your phone, you can dial telephone numbers directly from the keypad of your automatic dialer. Simply press the C/CE button to clear any number on display. Lift the handset of your phone and dial the number directly from the Micro-Dialer. The number you dial will be on display and remain displayed until you clear it using the C/CE button again.

Since the Micro-Dialer displays only sixteen of the thirty-two memory stored numbers, it utilizes a two-position grid for locating and storing the first sixteen numbers, followed by a second group of sixteen.

To store numbers (Fig. 5-23) press the storage key (St). Select any of the thirty-two storage locations. With the Memory Locator Grid in the downward position, numbers one thru sixteen are stored. Press the key that identifies the location, the Memory Locator Key. In sequence, enter the desired phone number by pressing the number keys. If it's a long distance number, remember to push the number one first, then the area code, followed by the seven digit number. Press the storage release key (StR). The number is now stored in memory.

For the second batch of sixteen numbers, gently slide the Memory Grid upwards. The Pause key (Pa) and Pause Release Key (PaR) are used when the Micro-Dialer is located in an office where calls can only be made through a switchboard.

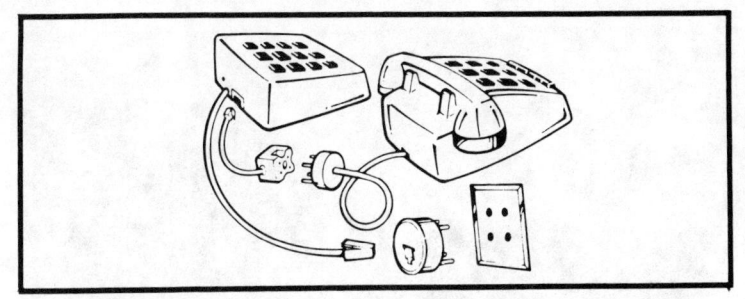

Fig. 5-22. A four-pin adapter connecting Micro-Dialer to your phone jack.

Whenever you wish to change a stored number, simply reenter a complete new number in the same manner as described. The Micro-Dialer automatically eliminates the old number and replaces it with the new one.

Automatic dialing is accomplished by picking up the receiver of your phone and pressing the desired Memory Location Key. The number will be dialed automatically.

You can also use the Micro-Dialer as a five function calculator. Press the Calculator Key (Ca.) The letter C will appear in the readout display. You can now use the Micro-Dialer as you would any desktop calculator. To return the unit to the automatic dialing mode, simply depress Ca again.

Maxxima Electronics Corporation supplies the Ringe-Dinge Number Storing Device that lets you record, store and automatically dial up to twenty-one separate phone numbers. Advanced circuitry permits pushbutton dialing, even on rotary dialing lines. A special

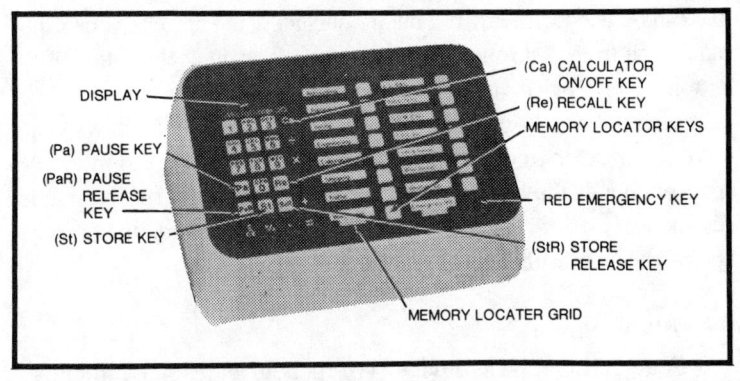

Fig. 5-23. Keyboard of the Micro-Dialer.

Fig. 5-24. Fone • A • Lert extension ringer (Courtesy of Floyd Bell Associates).

"last-number-dialed" feature automatically records any number you manually dialed for immediate recall in case of a busy signal. Color coded key caps can be placed over any special purpose buttons, such as fire or police department numbers.

Alaron Inc. also offers a number of automatic dialers. One unit has a thirty-two number memory capacity, while their less expensive device has a capacity of sixteen numbers. Both dialers work with any rotary or pushbutton phone.

Technology Applications Corporation has an automatic dialer that can be used with any telephone, including decorator phones. It can switch from dial to pushbutton and back again at the touch of a button. The Device has a twenty number memory that can be increased to forty by plugging two Rapidials together. It allows you to verify or refer to any numbers via an LED display. An internal speaker sounds the busy signal before you pick up the receiver and automatically cancels the call if you, the one being called, do not pick up the receiver within thirty seconds.

Fone•A•Lert

Fone•A•Lert is a device 4¼ inches wide × 3¾ inches high × 2½ inches deep that lets you hear your phone when you are some distance away from it (Fig. 5-24). It produces a piercing, audible signal synchronized to the ringing of your phone. There are no electrical connections to make, there is no dismantling of your phone, there is not even anything to plug in. The unit operates on a regular 9V transistor battery. Connection to the phone is made by means of a suction cup that is placed on the outside of the phone in the area of the bell. The Fone•A•Lert comes with forty feet of cable, and if you desire, you can add any length of 24 AWG, two-conductor speaker wire to it.

Bel-Ringer[R]

The same company that supplies the Fone•A•Lert also supplies an electronic telephone ringer (Fig. 5-25). Many companies supplying decorator phones may have this device already installed. In case you have purchased a telephone with an ordinary ringer, you can install the Bel-Ringer in the phone yourself—*not* in a phone company supplied telephone. Be aware, however, that any device you use with your telephone has to be FCC approved and has to meet the conditions of your local central office. Check with this office before you attempt any changes in your telephone.

Fig. 5-25. Bel-Ringer electronic bell for your phone (Courtesy of Floyd Bell Associates).

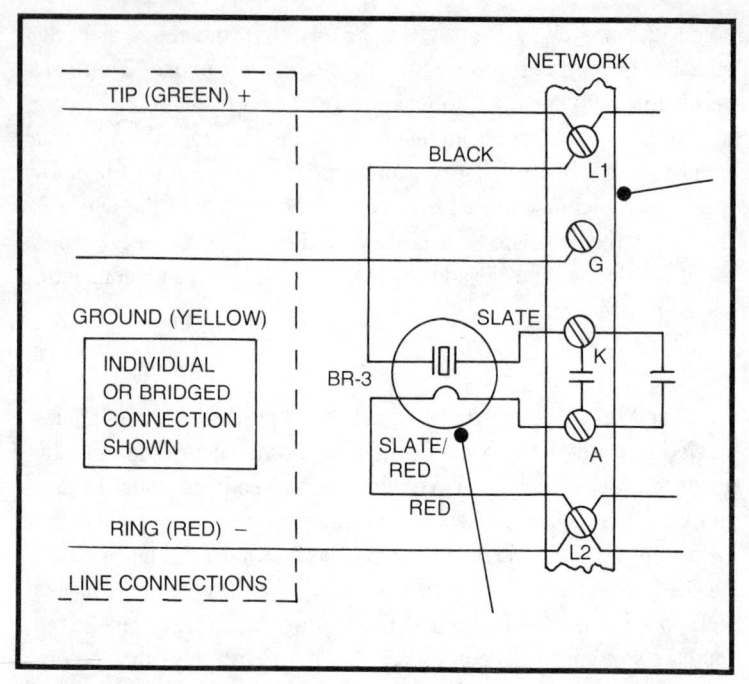

Fig. 5-26. Attaching the Bel-Ringer to the telephone.

The Bel-Ringer requires internal connections to the telephone which necessitates you having to open the case of your telephone.

Phone Silencer

Another device that requires a connection to the internal wiring system of the telephone set is the Phone Silencer by Zoom Electronics (Fig. 5-27).

It clips to the bottom of your phone and prevents unwanted phone calls. Instead of placing the phone off the hook or disconnecting the plug at the jack, this device eliminates the ringing sound of your phone while it gives a "ringing mode" to the caller. In other words, your caller hears a ringing tone, but you do not.

Again, if you want to attach the Silencer to your set, the decorator phone or other type telephone you have purchased yourself, you have to remove the phone's housing, disconnect certain wires at the terminal (Fig. 5-28) and connect the Silencer to the system (Fig. 5-29).

Pocket Yellow Pages

A handy pocket yellow pages manufactured by Canon is a combination calculator and telephone directory memory. The unit stores twenty of your most frequently dialed numbers in its memory and let's you recall them simply by entering the person's name or initials.

The keyboard has letters and numbers like the touch-tone pad of your phone. Want to call Alice? You enter A L I C E and the display shows Alice's phone number.

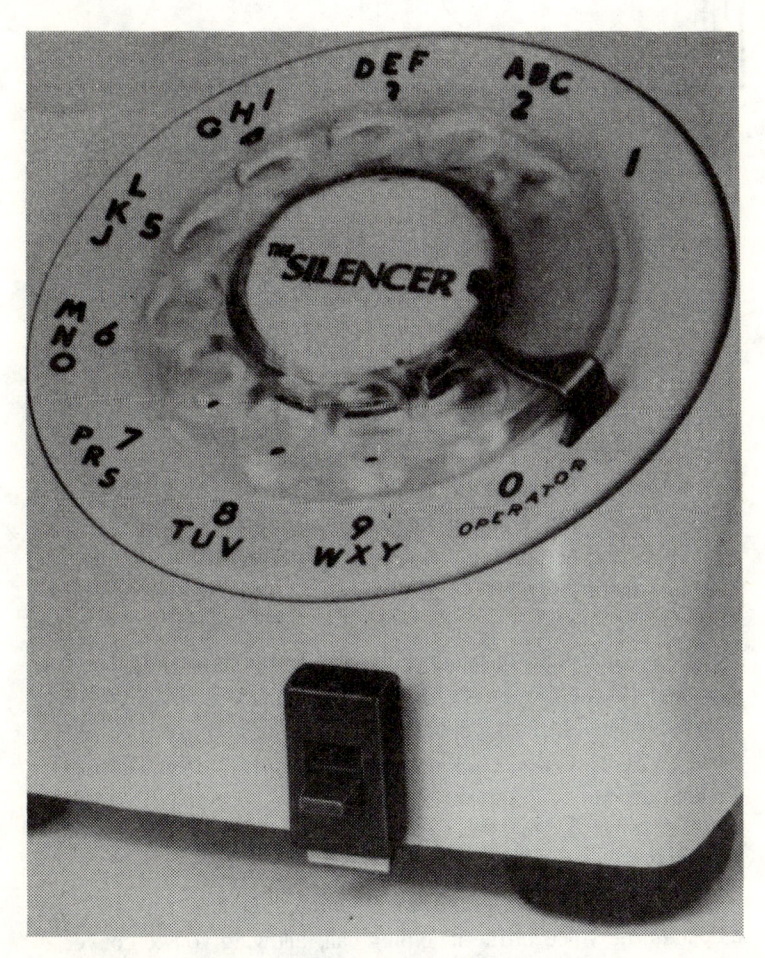

Fig. 5-27. The telephone silencer (Courtesy of Zoom Electronics).

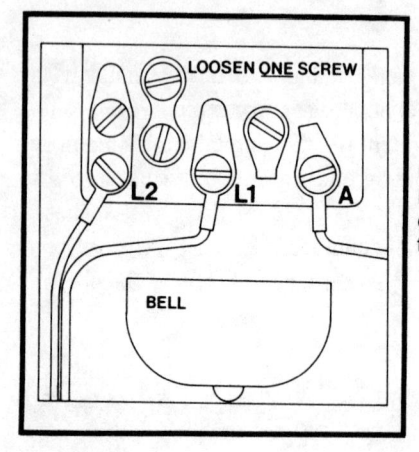

Fig. 5-28. The red wire to connection L1 or L2 of your phone has to be disconnected and reconnected to the lug of the Silencer.

Gobbler

Gimix Inc. has available a "Gobbler," a device that replaces the present bell of your phone. When your phone rings, you hear a turkey gobbling instead of your bell.

Hellos

Communico offers you prerecorded messages in case you're tired of using your own voice for your answering recorder. Or, perhaps your friends and others feel that another voice would be an improvement.

Communico has many volumes of "Hellos" available. They contain voice imitations of famous movie stars, recording personalities, and humorous characters. These are quality studio recordings made by some of Hollywood's top talent, with original music and sound effects.

The same kind of product is offered by Namedroppers and Phonies. Phonies' cassette tapes carry humorous impressions of thirty-five well-known public figures. Among those are: W.C. Fields, Groucho, Bogart, Woody Allen, Jimmy Carter, Kissinger, Cosell, and Muhammed Ali.

Amplifier

Panasonic is marketing a telephone amplifier, a unit that amplifies the voice of the incoming call, allowing you to converse without having to hold the handset to your ear.

Recorder

Saxton Products has the Tele-Beeper, a fully transistorized automatic telephone recorder. The Beeper is activated when you lift the handset from the hookswitch. The device also automatically turns the tape recorder on and pulses a "beep" signal into the conversation which is being recorded. Tele-Tender Systems has a similar unit available. The Tele-Recorder Device automatically records incoming and outgoing calls.

TELEPHONE ANSWERING MACHINES

Phone answering machines are the most widely used and known telephone accessories in existence. They come in a variety of styles, shapes and function. Since it is impossible, within the framework of this book, to describe all units that are on the market, I'll limit myself to the description of a number of well-known ones.

Ford Industries manufactures the Code-A-Phone Series, the 1200 and 1400 offering miscellaneous features such as variable announcement capabilities and message capabilities. The Model 1600 is a fully integrated telephone set-answering machine with

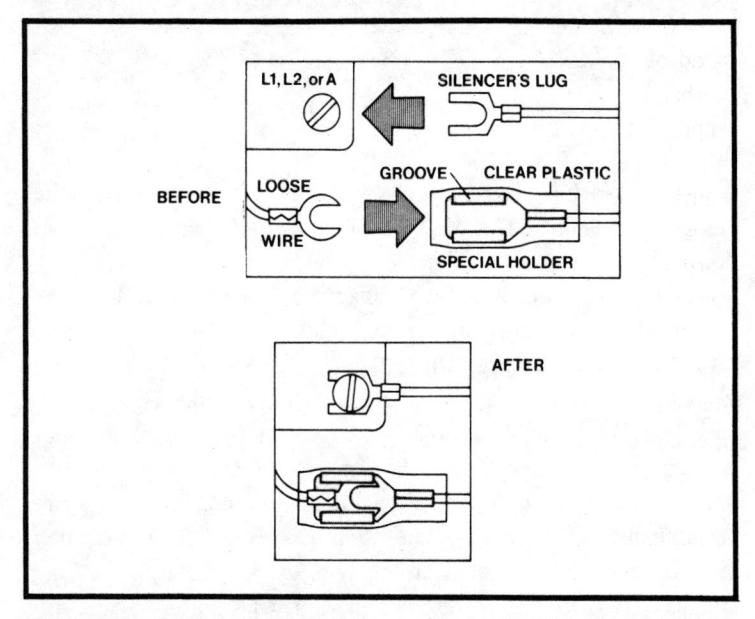

Fig. 5-29. The disconnected red wire is connected to the Silencer's lug, and the other lug of the Silencer is connected to L1 or L2.

Fig. 5-30. Fully integrated, decorator answering phone (Courtesy of Ford Industries).

which you can make outgoing calls while the machine remains in the answering position. The Model 1500 (Fig. 5-30) adds a touch of decorator elegance to your home.

TAD Avanti Corporation manufactures the RecordaCall Series of answering machines Models 60, 70 and 80.

Phone-mate supplies three different types of answering devices the Model 4000, C-Vox 8000, and Remote 9000 (Fig. 5-31).

The Remote 9000 records only as long as your caller talks, but you can set a time limit of up to three minutes. Your caller has plenty of time to leave a lengthy message. A six second pause causes the unit to hang up, ready for the next call. Controlled voice activation prevents any caller from deliberately tying up your line by putting his phone next to a radio, TV, or tape recorder. Only his voice can keep the unit going. The VOX activation also allows the unit to store long or short messages, one after the other, without dead space between each message.

The Audio Scan system allows you to hear what's on the tape while you rewind: how many messages, how long each one is, whether or not there is voice present, and where to easily locate a message you wish to repeat.

With the Remote Call Pickup, you can call in from any telephone, anywhere, anytime. Simply sound your coded pocket tone key and hear the messages played in complete privacy over the phone.

With the remote 9000, both your answering machine and your telephone have to be connected to the telephone network. If you have two phone jacks on the same number, you can plug the unit into one, and the phone into the other. However, in case you have only one telephone jack, or, if for greater convenience you want to connect your phone plus answering unit to the same modular jack, you can use an adapter for this type of connection (Fig. 5-32 and 5-33).

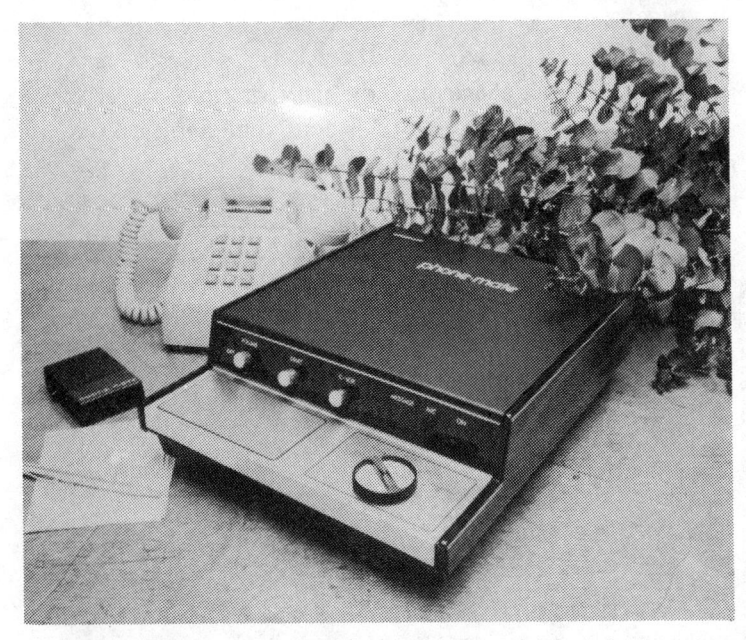

Fig. 5-31. Telephone answering machine with voice activation and scan system (Courtesy of Phone-mate, Inc.).

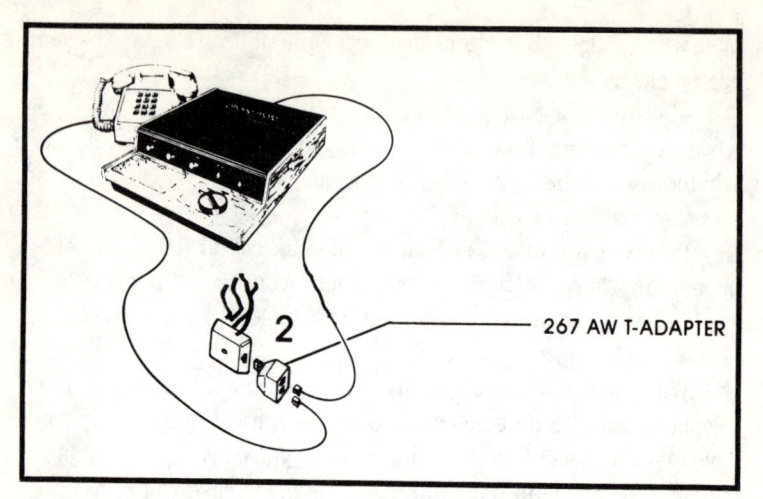

Fig. 5-32. Using a 267AQ adapter with a modular jack.

Notify your phone company about your installation and advise them of the FCC registration number of your answering machine, its ringer equivalence, and its protective circuitry number.

The Phone-mate 9000 answerer is basically two cassette tape recorders connected and controlled by switching similar to computer circuitry. When your phone rings, the answering machine responds by seizing the telephone line and delivering your prerecorded message to the caller. This prerecorded message is recirculating on an endless cassette cartridge. At the end of your message, a tone signal

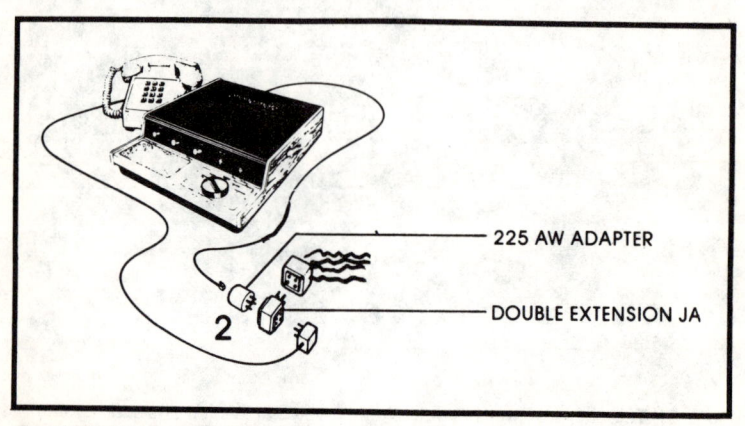

Fig. 5-33. Using a 225AW adapter with a four-pin outlet.

automatically switches the circuitry to the second tape to record your caller's message.

There are a number of ways to program your tape, but never say "I'm not in now...," or "We're on vacation...." The caller never has to know whether you're there or not. An ideal outgoing message is: "Hello! This is the _____ residence. Your call is being answered by our (name of unit). We would like to talk to you, so stay on the line. After you hear the tone, just leave your name and phone number and we'll call you as soon as we can. You can also leave a brief message. But be sure to leave at least your name and number so we can return your call. Thanks for calling."

Operating instructions for the Phone-mate Remote 9000 (Fig. 5-34) follow. To record your message:

- Insert both cassettes to lay flat when in position. Right-hand cassette must have full reel at top.
- Plug unit into regular 115V outlet.
- Turn unit *on* by rotating *volume control* clockwise. Green power *on* indicator will light. The left-hand outgoing announcement cassette will automatically cycle once to ensure that the tape is properly positioned to record your announcement. Wait for it to shut off.

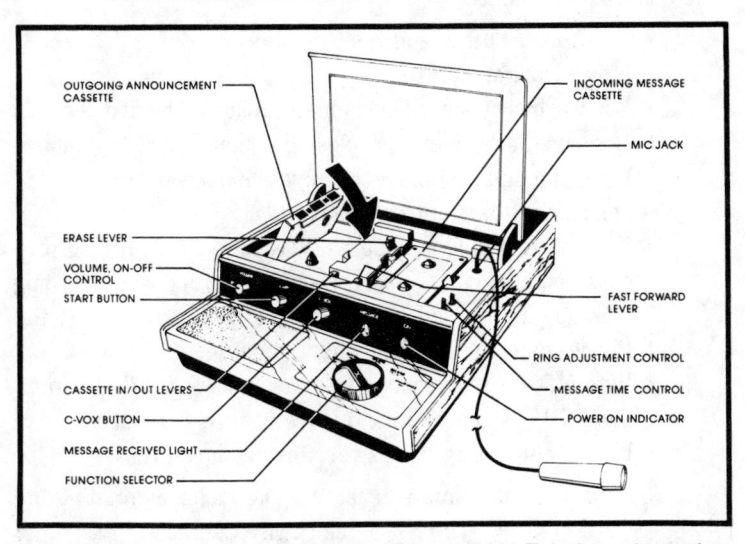

Fig. 5-34. Illustration of the Phone-mate Remote 9000 Telephone Answering Machine.

- Turn *function selector* to *record announcement*.
- Plug in microphone. Hold mike about eight to nine inches from your mouth and speak in a firm clear voice.
- Press and hold in *start* button. Wait two seconds, then begin announcement holding in button as you record.
- Release *start* button when announcement is completed and wait five seconds by slow count.
- Immediately press *start* button again and hold it in as you record this closing statement: "The time for leaving your message has expired. Please hang up now." Repeat this over and over, with no pauses between repetitions, until the outgoing announcement cassette completes it cycle and stops.
- Unplug mike.
- Turn function selector to *record calls*.

NOTE: It is important to accurately record the closing announcement. Filling the outgoing announcement tape with your voice ensures proper functioning of your unit when you use your Pocket Tone Key for remote call pickup of your incoming messages.

Having recorded your outgoing message, test it according to the following procedure:

- Press *start* button and release.
- Adjust volume control.
- You will hear your announcement, followed by five seconds of silence, and then your closing announcement.
- When the tape completes one cycle and has returned to the beginning, it will automatically stop.
- If reprogramming is necessary, repeat the recording process as outlined previously. Having recorded your outgoing message and tested them, you are now ready to set the Phone-mate to record incoming messages.
- With unit *on* and function selector in *record calls* position, push *C-VOX* button in and leave it in this position.
- Unit is now ready to answer your incoming calls.

You can adjust the time you allot to each caller by rotating the *message time control*. Turning it clockwise increases the time; counter-clockwise reduces the time. If the caller speaks for only ten to twenty seconds, the unit waits for six to eight seconds after he stops talking, and then hangs up, ready for the next call.

When you wish to pick up messages by remote control, you'll have to leave the unit in C-VOX control. This also prevents any caller from tying up your answering unit.

You can set the Phone-mate 9000 to answer on any ring you wish, from one to six. The advantage of setting the unit on the third of fourth ring is that you don't have to remember to turn your machine on when you leave. It is always on. If you're in, you can pick the phone up at the second ring, before the unit activates the answering mode.

To adjust the number of rings, turn *ring adjustment control* clockwise to increase the number or rings; counterclockwise to decrease this number. This control is not calibrated to indicate specific number or rings because of voltage variations in different parts of the country.

When you're at home, but busy, the monitor feature of the unit enables you to hear all your calls as they come in. This allows you to select the calls you wish to answer personally at that time. To answer incoming calls:

- Turn *volume control* to audible level.
- If you choose to talk to the caller, pick up the phone and turn *volume control* to *off*.
- Talk as long as you wish.
- At the conclusion of your conversation, turn *volume control* back to *on* position. Unit will automatically reset itself for automatic answering.

The *message* light will turn on when the unit has answered a call. If the light is on when you return, play back your incoming messages by:

- Turn *function selector* to *rewind*.
- Incoming message tape will rewind to its beginning.
- The Audio Scan feature will let you hear voices in high speed reverse, allowing you to determine where there are calls on the tape.
- Turn *function selector* to *playback* and listen to your messages.

- After you've listened to your calls, return *selector* to *record calls*, and unit is ready to answer new incoming calls.

In order to call into your unit to check on messages, always leave the unit in *C-VOX* mode.

- Call your phone from any phone, anywhere.
- As you're listening to your outgoing announcement, hold pocket tone key instrument against mouthpiece and sound tone for several seconds.
- Outgoing announcement will stop and you will hear incoming messages as the tape rewinds, in high speed tape reverse, like the sound of Donald Duck.
- Incoming message tape will now playback over the phone as you listen.
- When you've heard all the messages, place pocket tone key against mouthpiece again and sound tone. As you are listening to your own outgoing announcement, hang the phone up. The answering machine will complete cycle, hang up and be ready for the next incoming call.

When you are ready to clear the incoming message tape of old calls:

- Turn *function selector* to *rewind* and immediately flip *erase lever* into down position.
- When incoming tape stops, it will be completely erased of old messages. Turn *function selector* back to *record calls* and unit is ready to tape new messages.

There are quite a number of other brands of answering machines available. Answerex has two kinds on the market: both have a modular jack, dynamic mike, electronic erase, variable beep tone, a monitor, a two or four ring selector, and remote capability.

Panasonic is also in the answering machine market with the Receptionist. This unit has double cassettes, an anounce only function, a thirty or sixty time selector for incoming messages, and a recording feature for two-way conversations, and can be used as a dictating machine.

Quasar Microsystems' answering machine, Call Jotters, comes in three versions. The most sophisticated Jotter is a complete central communications facility which answers, announces, records, monitors, with remote playback and remote announcement change.

Sanyo Electric also has three answering machines on the market, with such features as a special loop tape for outgoing messages, built-in mike, monitor function, voice activation, time controller, auto stop, quick erase, and remote control.

NOTE: REMEMBER, ANY TIME YOU INSTALL OR USE A PHONE DEVICE THAT REQUIRES A CONNECTION TO THE EXISTING PHONE NETWORK, YOU SHOULD INFORM YOUR LOCAL CENTRAL OFFICE OF YOUR INTENT. GIVE THEM THE FCC REGISTRATION NUMBER OF THE DEVICE. THE AUTHOR AND PUBLISHER OF THIS BOOK CANNOT BE HELD RESPONSIBLE FOR ANYONE CONNECTING A DEVICE TO THE NETWORK BASED ON THE INFORMATION PRESENTED. IT IS UP TO THE USER TO CHECK WITH HIS TELEPHONE OFFICE! THIS BOOK IS WRITTEN IN ORDER TO INFORM YOU WHAT IS AVAILABLE ON THE MARKET BASED ON NEW TECHNOLOGIES, NOT TO INFORM YOU WHICH DEVICES ARE REGISTERED AND WHICH ARE NOT.

Decorator Telephones

The basic idea was a little crazy: redesign the telephone in sparkling plastic, gleaming with chrome trim. Then make it transparent, with the sprawl of wires, relays, and capacitors in plain view. And, as such ideas sometimes do, it caught on.

It takes imagination, creativity, and technical know-how to design a decorator phone.

An engineer and a designer for GTE who both contributed to the design of the GTE Starlite telephone attested to the work involved.

Every feature of the Starlite telephone (Fig. 6-1) was arrived at after considering a great many possible solutions, but the job is now completed and they are satisfied that the final design represents a unique combination of desired features. They said that the Starlite uses components of standard size and characteristics, which is one reason why they decided against a one-piece design combining handset, dial, hookswitch, and transmission components into a single unit.

The Starlite provides unexcelled performance from the standpoint of both transmission effectiveness and fidelity. This was another reason for choosing a two-piece design. The handset contains the transmitter and the receiver—nothing else. The user is not made to hold in his or her hand the extra weight of the remaining components. Consequently, it is not necessary to reduce the weight

Fig. 6-1. The Starlite[R] telephone (Courtesy of GTE Automatic Electric).

of these components by using less iron, or copper than optimum for most efficient transmission. The new transmission circuit uses varistors for automatic compensation of differences in telephone line characteristics.

The Starlite is light enough so that it may easily be lifted and moved about, yet heavy enough to stay where it's put. A rubberized cork pad covering the entire bottom of the base keeps the Starlite telephone from moving while a number is being dialed. And the user, after dialing a number, can lean back in his chair without fear of pulling the telephone off the desk. The Starlite has a new coiled cord which extends to almost five feet, yet exerts only a few ounces of pull on the telephone.

This telephone is not only light in weight, but also designed for easy lifting. Slightly projecting edges on the front and back of the nearly square dial plate provide a very efficient handhold, in contrast to the smooth plastic surfaces which ordinarily provide a somewhat less than certain grip.

The Starlite telephone has a unique lighted dial, using an electro-luminescent instead of the usual incandescent lamp. The electroluminescent lamp provides a very soft and pleasing greenish light over the entire face of the dial. The operating cost is amazingly low, a penny or two a year under most conditions. Also, it's an almost permanent light source, and will normally require no replacement or other maintenance for many years. The user is not burdened with the chore of making periodic replacements. The connection to 115V AC for the dial light is kept entirely separate from the telephone wiring and cord. The telephone user can adjust the brightness of the dial light to suit his or her personal preference. A serrated wheel in the front of the telephone gives fingertip control.

The Starlite telephone provides generally the same "walking handset" feature that has proved so popular in the Type 80 telephone, the standard phone. The design of the cups in the base and the sloping top of the housing make it very hard for anyone accidentally to miss the cradle while replacing the handset. Even if the handset is laid on its side in the cradle, it stays in place and depresses the hookswitch, hence the Starlite telephone should establish a particularly good record for reducing off-the-hook receiver trouble.

Along with the Starlite phone, GTE's engineer and designer worked on a small, attractive ringer box, approximately 5 inches × 5 inches × 2 inches for use with the Starlite telephone, or any other ringerless telephones. It is available in a full range of standard colors, and makes a pleasing combination for an extension located beyond the hearing range of the mainline ringer, or for use when the Starlite telephone is to be the mainline instrument.

GTE'S DECORATOR PHONES AND ACCESSORIES

This was of course some years ago. GTE Automatic Electric still has the Starlite[R] phone in its line, plus the Fashion Plate[R] phone, Styleline[R] phone (Fig. 6-2), Hands-Free Speakerphone (described earlier in this book), SoundBooster[R] phone, Two-Line Phone, and others. All of them have a choice of rotary dial or Touch-Tone[R] in a variety of colors, such as light blue, avocado, antique white, beige, tangerine, autumn gold, white, espresso brown, sand beige, and black.

Yes, the days of the old standard black phone are gone. Colors and shapes of the telephones are designed to match a variety of interior designs. GTE, for example, has made available a booklet "Design Trends: A Look at Today's Most Exciting Decorating Ideas," and of course, decorating ideas matched to the decorator phones.

Fig. 6-2. The Styleline^R telephone (Courtesy of GTE Automatic Electric).

The Code-Com is an aid to deaf, and deaf and blind persons which enables them to tap out messages over regular phone lines. The set, connected to a telephone, will allow a deaf person to "see" phone messages in coded flashes of light, or "feel" them in vibrations of a finger pad. A sending key, used like a telegraph key, is for deaf people without speech.

An external ringing bell aids those with impaired hearing. The bell can be placed anywhere, so you needn't be in the same room as the phone to hear the bell.

NOTE: The little^R that goes with the various names of the telephones means that that particular name is a trademark of the company supplying the phone. With the fierce competition among telephone companies and decorator phone suppliers, everyone is protecting their design and name with a trademark registration. For example, GTE's Styleline^R phone looks similar to ITT's Trendline^R (Fig. 6-7) and AT&T's Trimline^R, but there are subtle differences.

Fig. 6-3. The Snoopy and Woodstock Phone (Courtesy of AT&T).

AT&T'S DECORATOR PHONES AND ACCESSORIES

AT&T entered the design phone wave some time ago and they offer phones in five styles: standard desk, wall models, Princess and Trimline desk, and wall phones. These five basic styles have been adapted into a large number of special Designline[R] phones: Candlestick, Chestphone, Early American, and Sculptura, to name a few.

These new design phones can be a real extension of your personality. You can run the gamut from phones that look like food to phones that look like toys, antique phones to super futuristic, phones

Fig. 6-4. The Noteworthy Phone (Courtesy of AT&T).

Fig. 6-5. The Celebrity (Courtesy of AT&T).

made of mahogany and phones made of clear acrylic, even phones hidden in boxes so they don't look like phones.

AT&T's latest designs include The Snoopy & Woodstock Phone and the Noteworthy Phone (Fig. 6-4). The Celebrity (Fig. 6-5) is a design from a past era. It is available only with a rotary dial, but in two colors: blue accented with silver colored trim, or ivory accented with gold-colored trim. Figure 6-6 pictures the Elite Phone, a combination of simulated leather and glowing metal. The leather is dark green with gold colored trim and a dark green handset.

The TouchaMatic is useful for those who spend a lot of time on the telephone. It has a memory phone for numbers dialed frequently, with either sixteen or thirty-one memory buttons, record, and last number dialed buttons.

AT&T's Sculptura is the most unusual phone in the Designline. Although shaped like a bottom-heavy doughnut, the Sculptura offers a contemporary look, clean lines, streamlined design. Other Designline[R] options enable you to match your decorating scheme.

Notwithstanding decorative applications, AT&T also has special telephone handicap aids. Volume control handsets are designed for people with speech or hearing difficulties. Two types are available, one with a built-in amplifier that increases the volume of what is

Fig. 6-6. The Elite Phone (Courtesy of AT&T).

Fig. 6-7. ITT's Trendline[R] Telephone is similar to GTE's Styleline[R] and AT&T's Trimline [R].

heard, and one with an amplifier that increases the volume of what is spoken. The Telephone Adapter is for use with hearing aids. The Adapter solves difficulties with Trimline phones. It is an acoustic coupling device that is easily fitted and removed from the receiver, very small, and can be carried around so it can be fitted on any phone.

ITT'S DECORATOR PHONES

ITT offers a variety of decorator and functional telephones also (Fig. 6-8). Figure 6-8A is a phone-recorder combination. Classic Phones (Fig. 6-8B) comes in different styles and colors, such as walnut/brass, walnut/gold, and gold leaf/wood. The Chest Phone (Fig. 6-8C) comes in selected hardwood oak, or early American maple. Figure 6-8D is a telephone/radio combination. Figure 6-8E, DeltaPhone[R] comes in a white/gray or beige/green combination. The Multibutton Security Phone (Fig. 6-8F) puts you in touch with emergency services at the press of a button. The Gondola

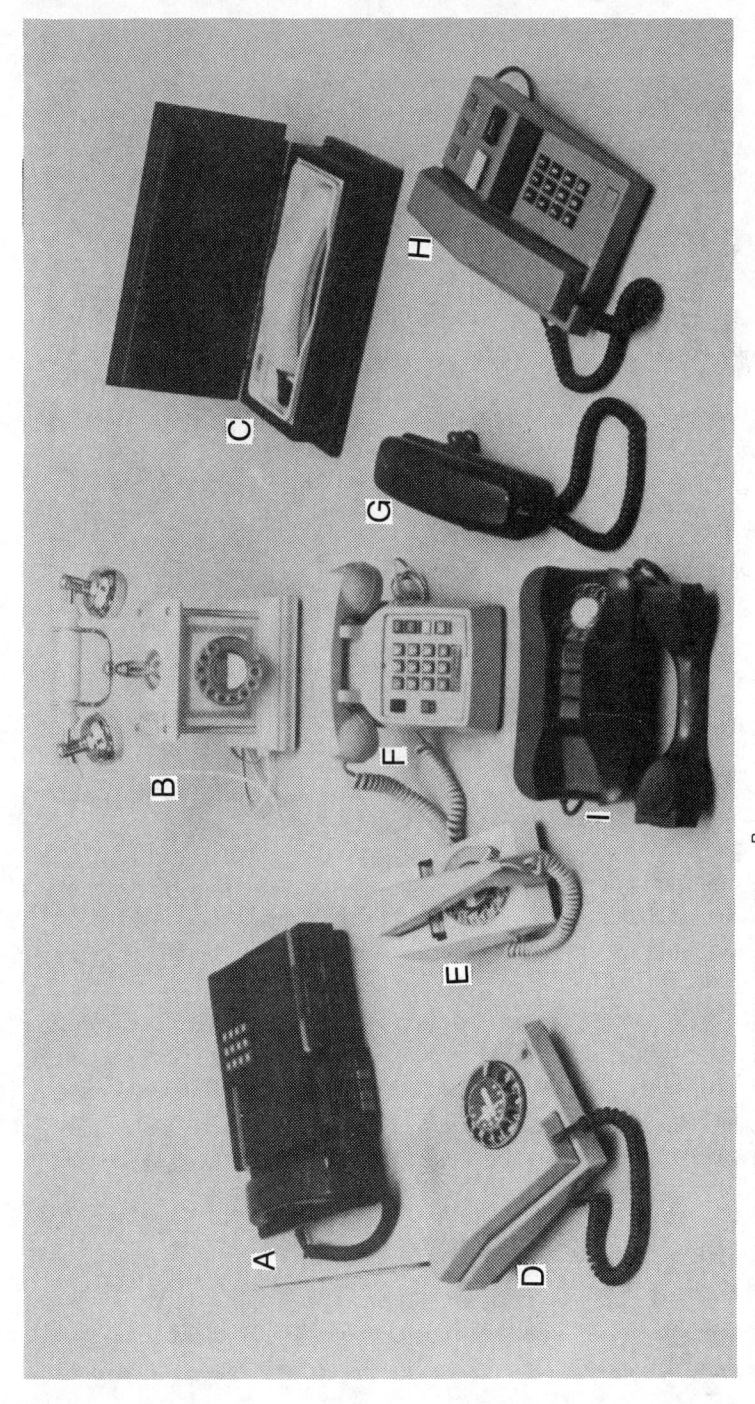

Fig. 6-8. A representative sample of ITT's Own-A-Phone[R].

Fig. 6-9. "School Desk" Phone (Courtesy of Telcer Telephone).

Trendline^R Phone (Fig. 6-8G) comes in a variety of colors. The Alarmclock Phone (Fig. 6-8H) is an alarm and wake-up clock/phone combination. Figure 6-8J, the JewelBox^R Phone, comes in beige plastic, or red, tan, or green leather.

ITT's latest telephones include the UltraPhone^R in a green leather/green phone combination, brown leather/ivory phone, or tan leather/beige phone combination; and the Domino^R—in a black/white combination. The ITT telephones are sold and marketed under the name Own-A-Phone^R (registered tradename).

OTHER DECORATOR PHONES

Tel-O-Fun^R, a division of Leever Bros. Company, markets decorator phones in prices ranging from $65-$160. Ericofon is a dial-in-base model. Contempra is a model with space-age styling.

The French Connections are four antique styled telephones, either gold-plated or with detailed antique brass. The Americana Group is comprised of replicas of classic upright phones from the roaring 20's and two other antique styled phones.

Carter-Craft, a division of Carter Corporation, offers a variety of decorator phones, unfortunately displayed in a black and white catalog.

An exciting array of decorator phones is offered by Telcer Telephone USA. All their telephones, thirty-three in all, vary in price from $220 to $490, and are exclusively in the antique style. Imported from Italy, their line is comprised of onyx and gold combinations, hand inlaid mahoganies, rosewoods and steamed beech, special lacquer crackle finished paints, and fine Italian and Moroccan leather, accented with gold foil detailing and 18K gold plate trim. Figure 6-9 shows a "School Desk" model in onyx green; Fig. 6-10 a gold lacquered phone called "Model Sweden."

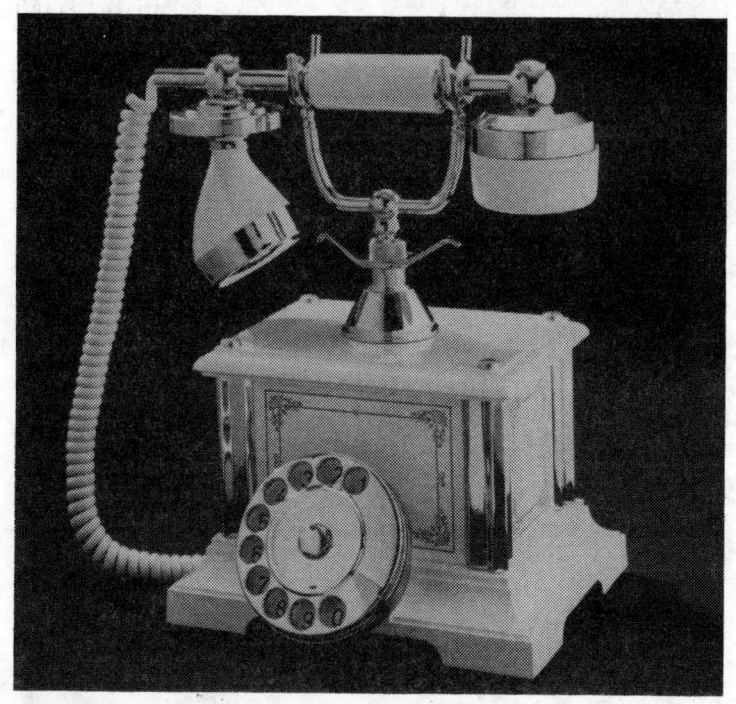

Fig. 6-10. Gold lacquered "Model Sweden" Phone (Courtesy of Telcer Telephone).

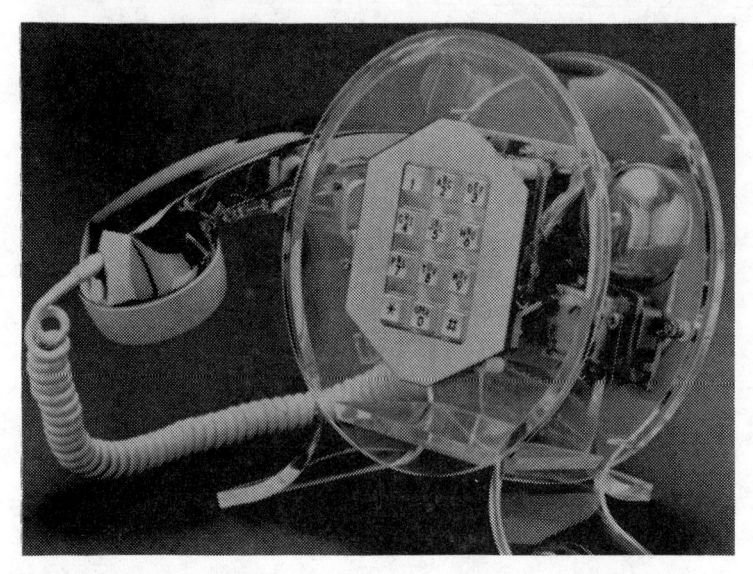

Fig. 6-11. The Disco Clear Phone (Courtesy of TelConcepts Inc.).

Fig. 6-12. The Chromefore Phone (Courtesy of Teleconcepts Inc.).

Fig. 6-13. Gold Arabesque Phone (Courtesy of Fold-A-Fone).

Equally exciting, but totally different in concept are the decorator phones from TeleConcepts Inc. All futuristic in design, their line consists of such phones as Shellamar Abalone Disco Clear (Fig. 6-11), Apollo, Diamente, Chromefone (Fig. 6-12), Tempo Pearl, and Teledome.

Fig. 6-14. Petite French Phone (Courtesy of Fold-A-Fone).

Fig. 6-15. Colonial Wall Phone (Courtesy of Fold-A-Fone).

Fig. 6-16. Teen Telephone (Courtesy of Fold-A-Tone).

Fold-A-Fone Inc. offers a variety of antique styled and futuristic telephones: French Cradle Telephone; Gold Arabesque (Fig. 6-13); Petite French (Fig. 6-14), made of molded plastic and solid brass or white with gold trim; Bonnie & Clyde; Colonial Wall Telephone with its wooden cabinet, concealed dial, and brass ringer bells for authenticity (Fig. 6-15); Emanuella; Erica; Chicco; and the Teen Telephone with white, sleek, simple lines, a bright orange receiver cradle and lime green dial plate (Fig. 6-16).

Fig. 6-17. Dawn (Courtesy of Northern Telecom).

Fig. 6-18. Kangaroo (Courtesy of Northern Telecom).

In a most exciting campaign of color brochures, descriptions and photography, Northern Telecom markets its Imagination[R] series of decorator telephones. Dawn (Fig. 6-17) is a sculptured design available in four colors. Kangaroo (Fig. 6-18) has a note pad

Fig. 6-19. Doodle (Courtesy of Northern Telecom).

and pouches for messages and reminders, available in six colors with the bottom in leatherette, corduroy, or denim. Doodle (Fig. 6-19) is a variant of the Kangaroo, available in six colors, with the notepad and a pencil, but without the pockets.

Who can resist this kind of appeal to decorator phones!

Telephone Security Devices

Recent history has brought widespread recognition of the prevalence of electronic eavesdropping, not only in government circles, but in private business and civilian life. Although it is perhaps unlikely that your telephone is wiretapped, there is a way to determine if it is or not. To make sure that your privacy on your telephone line is retained, a number of manufacturers are actively selling voice scramblers and other such devices that make your words unintelligible.

VOICE SCRAMBLING DEVICES

Mieco Corporation's Privacom P-25A (Fig. 7-1) makes conversation between you and others completely private, provided that all the callers are equipped with the unit. The phone scrambler is a multiple-coded, self-contained, readily portable instrument that converts normal speech into unintelligible gibberish by interchanging the high and low tones before transmission over the phone lines. A similar instrument at the other end of the line inverts the speech tones, thereby converting the gibberish into clear speech for the intended listener. The exact translation of the speech tones is controlled by a tone generator within the scrambler instrument, making possible a number of different scrambling codes by varying the pitch of the translation tone generator.

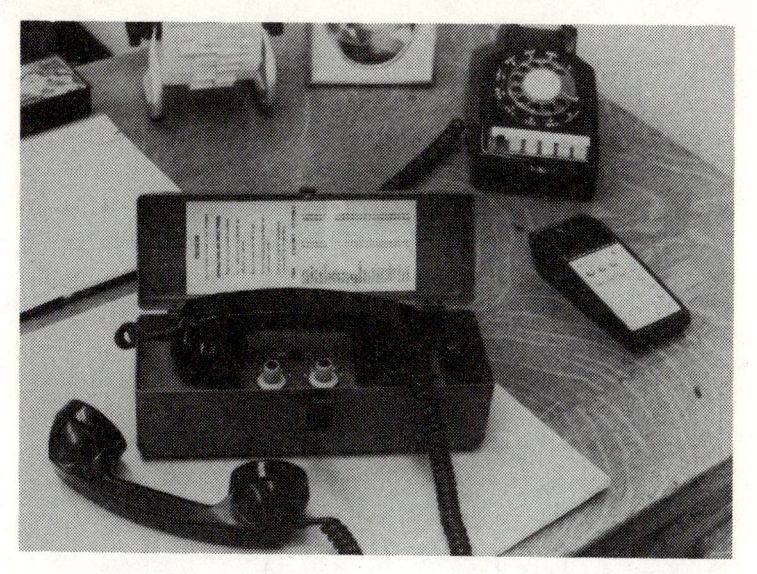

Fig. 7-1. Telephone Voice Scrambler (Courtesy of Micro Corp.).

Mieco Corporation's Code Phone provides individual speaking and receiving selection of twenty-five different translation codes on the control panel of the unit. The letter-coded switch controls the speaking translational tone, while the numerical coded switch controls the receiving tone translation. The alpha and numeric dials of the two Code Phone scramblers involved in the communication are set to complement each other, offering twenty-five different codes, according to Table 7-1.

Phone Scrambler can be put into operation by simply removing the handset of the scrambler and replacing it with the handset of your own phone, setting the code dials according to the code desired, moving the switch to *on*, and by carrying on the conversation using the scrambler handset.

A more complicated unit, although operating in the same fashion as the unit described above, is Datotek's DV-505. In this set, the clear voice input is split into five frequencies (bands), then rearranged (scrambled) into five output bands. In one of the security modes, the scrambling structure is changed each 0.25 seconds, thus rendering analytical attack impossible. Each new rearrangement is selected by a digital key generator using cryptographic techniques.

Table 7-1. Code Phone's Twenty-five Codes.

Code	Station 1	Station 2
Alpha	A1	A1
Bravo	B2	B2
Charlie	C3	C3
Delta	D4	D4
Echo	E5	E5
Fox Trot	B1	A2
Gold	B3	C2
Hotel	B4	D2
India	B5	E2
Juliett	C1	A3
Kilo	C2	B3
Lima	C4	D3
Mike	C5	E3
November	D1	A4
Oscar	D2	B4
Papa	D3	C4
Quebec	D5	E4
Romeo	E1	A5
Sierra	E2	B5
Tango	E3	C5
Uniform	E4	D5
Victor	B1	A2
Whiskey	A3	C1
Xray	A4	D1
Yankee	A5	E1

Over two million user keys (codes) are selectable on the front panel of the unit. An internal programmable plug increases the number of codes to thirty-two trillion (that's right—*trillion*).

Controllonics Corporation manufactures a unit based upon the same principles, except this scrambler looks like an ordinary telephone (Fig. 7-2). Operating on batteries, you place the handset of your regular phone on the base of the CT-300 scrambler and, after inserting a code key, you use the handset of the scrambler for your conversation. Their CT-300 is used for AT&T and ITT telephones, and the CT-301 for phones supplied by GTE and others.

Technical Communications Corporation has a number of voice scramblers available, plus a number of other communications security devices. Their Voice Privacy Device (Fig. 7-3) uses a wide deviation frequency displacement scrambling technique to provide a high degree of security which cannot be decoded by conventional techniques. Coding is accomplished in both the frequency and time domains. A unique frequency translation and variable time technique

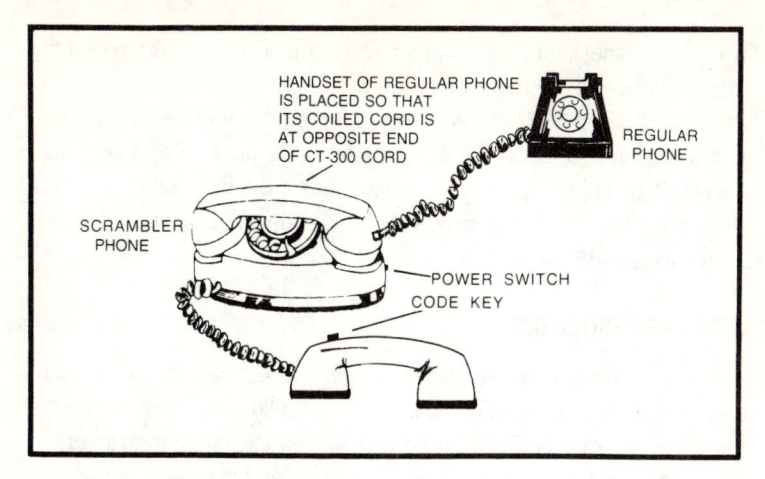

Fig. 7-2. Controllincs scrambler hook-up.

scrambles the voice under control of nonlinear digital codes. There are twenty-two frequencies selected in a pseudorandom manner. The transition from one frequency to another is not continuous, but uses a series selected from eighty-five frequency steps. There are eight sequences of 512 discrete frequencies each, stored in a memory. Sequences are selected at random, one at a time, under control of a pseudorandom code generator having almost 8.4 million states.

Frequency steps are changed every millisecond, and the duration of a sequence, while random, is always over 0.5 of a second. The code repetition cycle is over 1165 hours. There are almost eighteen billion codes in the system, of which over one million may be selected

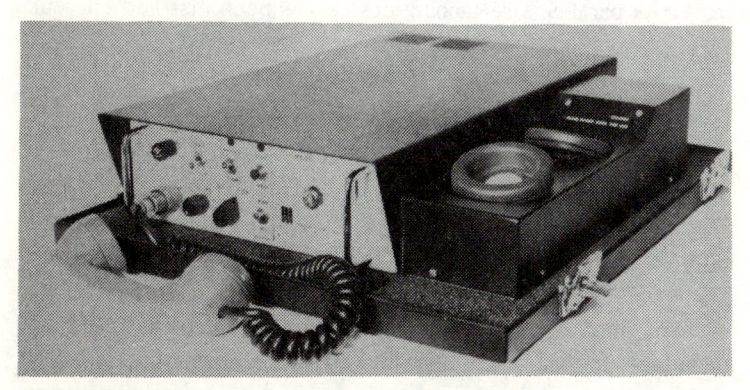

Fig. 7-3. Voice Privacy Device (Courtesy of Technical Communications Corp.).

131

from front panel switches. Additionally, a ten position internal switch increases user selectivity of codes to over ten million.

Synchronization between two units is accomplished by transmission of a 1.02 second long eight bit code preamble. The sync burst will be rejected by the receiving unit if the internal sync code does not agree with the received sync code. The sync code and scrambling code are not the same.

WIRE TAP DEBUGGERS

There are a number of devices on the market that give you around-the-clock protection against the installation of wire taps and telephone operated room bugs on your telephone line or instrument. Although wire tapping is in some cases legal and in other cases not, the use of a telephone tamper alarm will give you the security of knowing when your phone is being tampered with.

Law Enforcement Associates Inc. has a unit on the market called Tap-Alert (Fig. 7-4). It's a highly sensitive electronic device that when properly installed will give you twenty-four hour protection against attempts to tamper with your phone or telephone line. The unit is a voltage-controlled "switch" which will warn you that one or more of the following conditions have occurred:

- The testing of a wire tapping device has taken place, after installation of this device on your phone line.
- Your phone line has been temporarily cut to facilitate the installation of a series wire tapping device.
- A parallel wire tapping device has been installed on your phone line, effecting its voltage characteristics by 10% or more.
- A tone activated wire tapping device (harmonica bug) has been activated on your phone line.
- Someone has used or is using a telephone connected to your line.
- There has been an abnormal interruption of your telephone service.

You may, if you wish, connect an inexpensive cassette tape recorder to the Tap-Alert, allowing you to make a permanent record of the cause of any of these conditions. Be aware, however, that

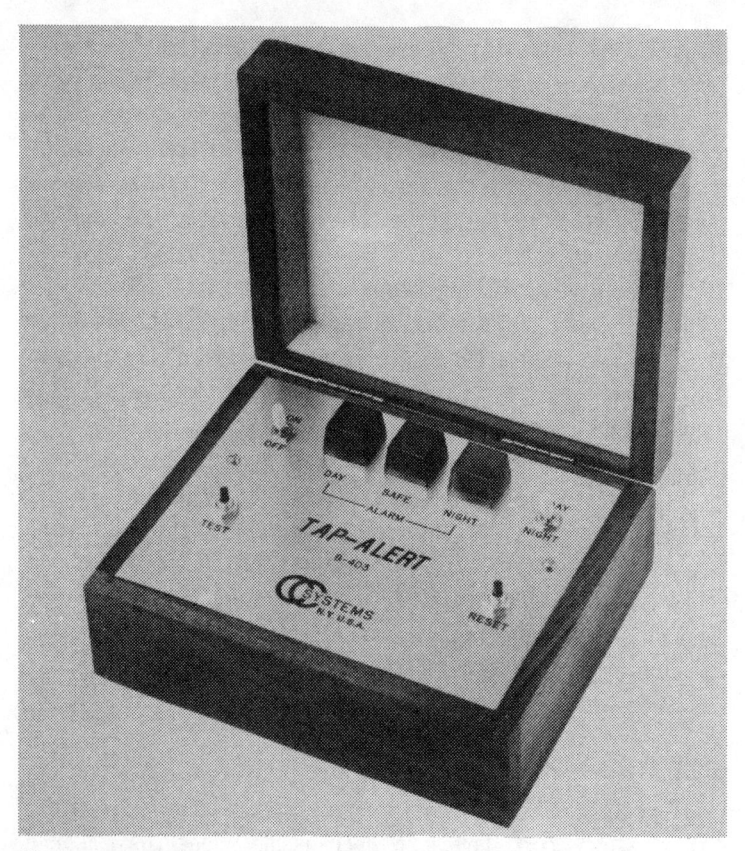

Fig. 7-4. Tap-Alert (Courtesy of Law Enforcement Associates, Inc.).

Federal law requires that one party to a telephone conversation be apprised that the conversation is being monitored.

Installation of the Tap-Alert is simple. Plug the brown cord from the unit to a 115V wall output. Plug the two grey cables into the mike and remote jacks of your tape recorder. Connnect the thin silver and gold wires from the device across the telephone line wires.

Remove the small telephone box cover mounted on the wall that your phone is wired into. Locate the red and green wires in the box. Connect one of the wires of the Tap-Alert, for example the silver one, to the screw terminal where the red wire is connected, and the other wire, the gold one, to the screw terminal where the green wire is connected. Flip the *day/night* switch on the front panel to the *day* position. Locate the sensitivity control on the rear of your unit, and using a small screwdriver, turn this control fully clockwise. Flip the *on/off* switch to the *on* position. Set the controls of your recorder as you normally would to make a recording.

The red *day* light should be lit, and the recorder should be running. Slowly turn the sensitivity control counter-clockwise a few degrees at the time until the red *day* light is extinguished and the green *save* light is lit.

If the above steps have been properly executed, the red light will be lit and the tape recorder will run whenever the phone is lifted from the hook. When you hang the phone up after your conversation, the unit will return to the green light and the recorder will stop running, unless a "tamper" condition exists.

During the evening, or when you're away from the telephone, the device should be set to the *night* position. The operation of the unit in this mode is identical to that of the day, except that when a tamper condition occurs, the yellow *night* light will light and stay lit until you push the *reset* button. This way you're instantly warned upon your return that a tamper condition has taken place.

The *test* button provides you with a "tamper," or "off-hook" simulation without the need of lifting the receiver of your phone.

VOICE STRESS ANALYZER SYSTEM

An exciting, new, and controversial product is the Voice Stress Analyzer (Fig. 7-5). This unit electronically measures inaudible microtremors in the voice, thereby determining the presence of stress. If the right questions are asked, stress can indicate lying.

The Voice Analyzer, manufactured by Communication Control Systems, can be connected to the telephone through an adapter and provides an exact numerical measurement of psychological stress in speech by electronically detecting the slightest subaudible modulations in the voice. The digital readout is continuous during the interview, so any variations in a person's normal voice pattern are reflected instantaneously in numerical jumps. No physical attachment to the speaker is needed. The entire system is contained in a lightweight, portable attache case.

As the person interviewed talks, the modulations in his or her voice appear as a digital readout on the screen. By querying the person initially with questions that can be easily answered truthfully,

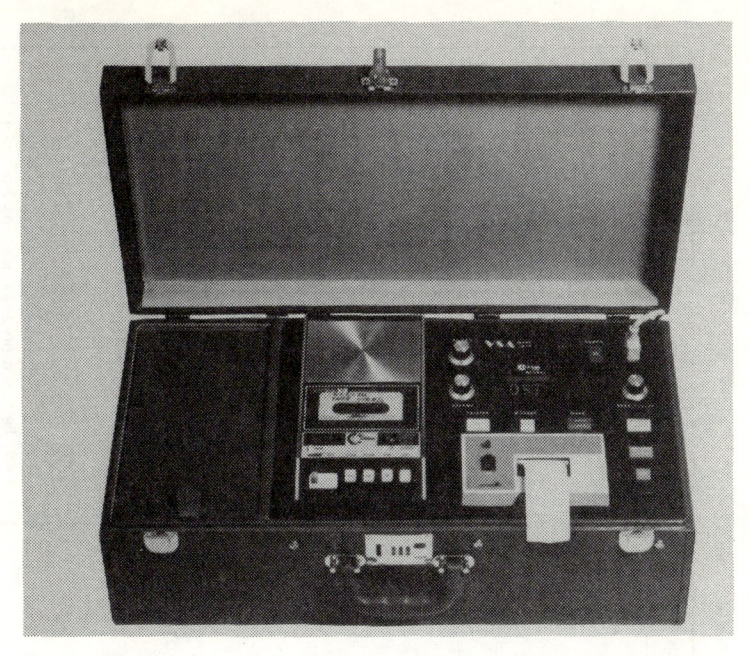

Fig. 7-5. Voice Stress Analyzer (Courtesy of Communication Control Systems, Inc.).

the normal range of his or her voice is established. In the digital readout, the normal average threshold is twenty-four. When answers to critical questions prompt higher-than-normal numbers on the screen, the person's stress immediately is apparent to the listener (Table 7-2). The numbers represent a sample interview of an actual suspect for theft. Actual numbers will vary according to the norm established in an individual interview.

Law Enforcement Associates Inc. manufactures a similar device, the Mark II Voice Analyzer, in a portable version and table top version.

Although this book does not deal with the pros and cons of the various devices on the market, I feel that a short analysis on this controversial subject of "lie detecting" is necessary, especially since the voice analyzers can be connected to the telephone, the person interviewed is not present at the analysis, and hence could conceivably be unaware of the analysis at all. Among the proponents of the technique are policemen and industrialists. Opponents include scien-

Table 7-2. Voice Stress Analysis.

Number in left column is keyed to the questions.	Numbers in right column refelcts the answers.	
1	27	Is your last name ____? Yes
2	21	Are we in New York City now? Yes
3	24	Is today Monday? Yes
4	86	Did you steal the $1250 in question? No
5	22	Did you go to school? Yes
6	83	Did you help or permit someone to steal the $1250? No
7	25	Do they call you____? Yes
8	7	Do you suspect someone of stealing the $1250? No
9	79	Do you know who stole the $1250? No
10	85	Did you steal the $1250? No
11	77	Did you lie to me on any questions I've asked? No

tists and engineers. However, supposedly physicians, lawyers, psychiatrists and a host of other people are very intrigued with these devices. The units sell for approximately $5000, but ads for them appear in consumer magazines, despite their cost, so obviously these devices have generated a lot of interest.

The controversy centers around two points: first, the need for experienced interrogators, and second, whether or not there are microtremors in the voice, and if so, whether or not they dissipate under stress. (There are also indications that stress is not necessarily related to lying and is often deeply rooted in circumstances that have no relation to the questions being asked.)

SECURITY ALARM SYSTEMS

Figure 7-6 illustrates the basic security alarm system and the components which make it up. Each component provides a vital function necessary for the proper operation of your alarm system.

For extra protection, an alarm system can be powered by using a power supply and/or battery. The power supply connected to a

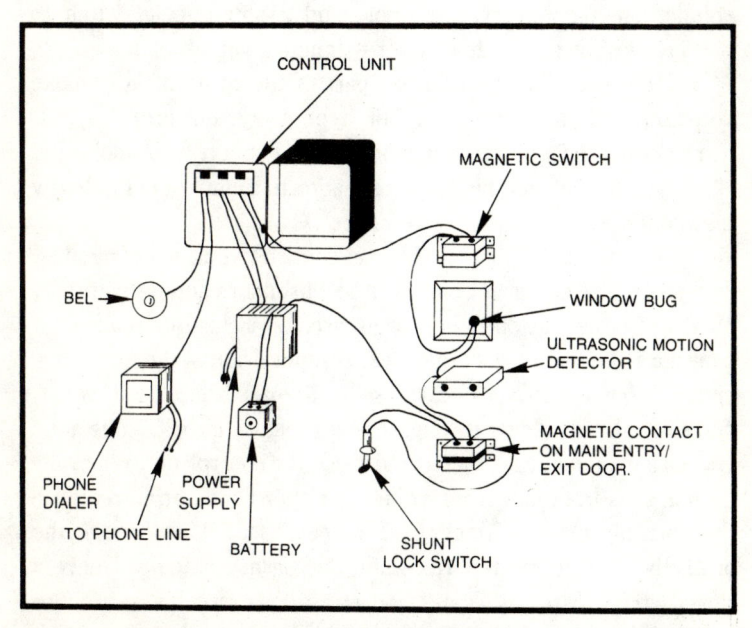

Fig. 7-6. Security alarm system wiring diagram (Courtesy of Emel Electronics).

regular 115V AC outlet reduces this current to low voltages for powering the whole system. Should the house current be turned off or become inoperable, the battery will automatically power the system.

The switches sensors and detectors (Fig. 7-7A) may consist of door and window switches, photoelectric detectors, ultrasonic and microwave detectors, floor mats, or smoke and fire detectors. All these components detect an emergency situation such as fire, heat, smoke, or a burglar. The detectors provide an output which is wired to a Control Unit.

The Control Section (Fig. 7-7B) encompasses the control unit and the power supply. The control section latches in an alarm state when tripped by any of the detectors. When the control section is in the alarm state, it connects power to a bell or siren. It can close a switch which operates 115V AC flood lights. It can also activate an automatic telephone dialer, a device notifying people via the telephone line of an emergency.

There is a magnetic switch for doors and windows, a window bug for glass breakage, an ultrasonic motion detector, a magnetic contact on main entry/exit door, and a shunt lock switch to activate/deactivate the front door magnetic switch.

Not shown, but available as part of the system, are smoke detectors and panic switches. This type of system protects your home or establishment from any intrusion, danger, fire/smoke and allows you to transmit this danger situation to a number of locales by means of your automatic phone dialer.

A control unit (in Fig. 7-7B) can contain two types of circuits. The protective circuit is connected to all sensors and detectors by means of either an open or closed circuit. A closed circuit is far superior to an open circuit: it provides supervision of the protective circuit. A closed circuit is called *series wiring*. In a closed circuit, current normally flows through the protective circuit, constantly energizing a sensitive relay located inside the control unit. When any switch, sensor, or detector is tripped, an open condition is created in the normally closed circuit. Current ceases to flow through the protective circuit and this results in the alarm sounding. Hence a closed circuit system will detect if the protective circuit wire is cut or broken, since current will not flow if this happens. An open circuit

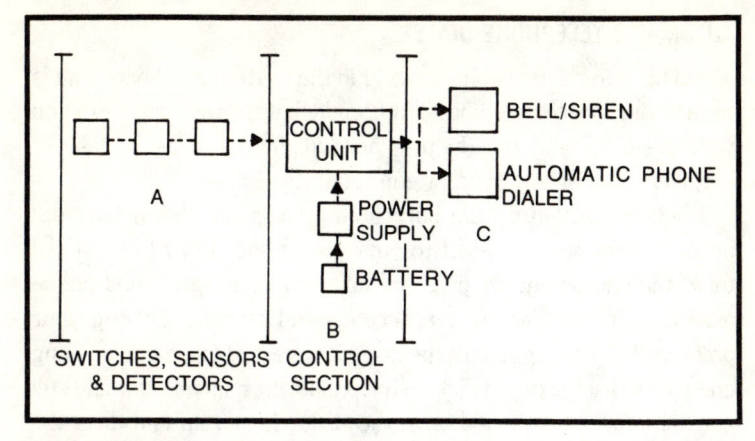

Fig. 7-7. Block diagram of a basic security alarm system: A, switch, sensing, and detecting section; B, control section; C, sounding and reporting section (Courtesy of Emel Electronics).

will not detect this type of malfunction. Closed circuit systems, called supervised systems, are used for almost all burglar alarm installations.

The sounding and reporting section (Fig. 7-7C) is actually the output of the control unit and is connected to bells, sirens, strobe lights, automatic telephone dialers, or virtually any warning device.

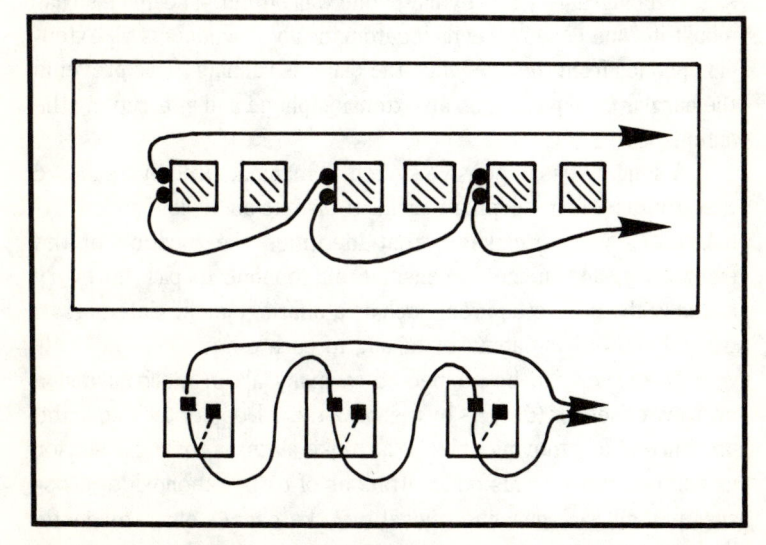

Fig. 7-8. Closed protective circuit for use with closed circuit control units.

139

AUTOMATIC TELEPHONE DIALER

The automatic telephone dialer in the system is a device that is used to alert police, neighbors, or friends that an emergency exits on your premises and that help is needed. This can be in case of burglary, fire, or any type of emergency.

When signaled by the burglar, fire, or panic alarm, the telephone dialer connects itself to your phone line. It waits for a dial tone, then automatically dials the first of four programmed phone numbers. It waits about five seconds and starts delivering your prerecorded message. This message is repeated three times—long enough so that party can pick up the phone after seven rings and still get full information,—and then the dialer hangs up and dials the second number. It gives the message and continues until all four programmed telephone numbers have been dialed.

An automatic phone dialer such as the one shown in Fig. 7-9, can have two channels, channel A for burglary, and channel B for fire. Each channel operates independently of the other and may dial different numbers and deliver a different message.

The dialer is connected to a regular 115V AC outlet, but it is also supplied with batteries in case the electrical current is cut off. The unit uses an eight-track tape cassette for prerecording messages. It plugs into your regular phone wall output. The unit also has a built-in "line seizure," which automatically disconnects all extension phones from the line when the dialer is running. This prevents the burglar from picking up an extension phone and interrupting the automatic call.

A similar, less expensive system is manufactured by Seaboard Electronics. Their telephone dialer transmits an audible coded signal, not a voice message, that identifies the location of the emergency, and utilizes a sensitive microphone to pick up every sound in the protected area, such as a buzzing smoke detector, or trespassers moving about or talking to each other.

The dialer can be connected to many alarm mechanisms: a microwave sensor (detects by means of radar Doppler technique the presence of any moving object), a smoke alarm, a toxic gas sensor (sensitive to dangerous concentrations of carbon monoxide, propane, natural gas, gasoline, diesel fuel, kerosene, etc.), magnetic door and window contacts, a freeze sensor (senses the existence of

Fig. 7-9. Automatic Telephone Dialer used in the alarm system shown in Fig. 7-6. (Courtesy of Napco Security Systems, Inc.).

cold air in case the heater goes off when you're away), a refrigerator sensor (a high temperature sensor which activates the phone dialer when the temperature in your refrigerator/freezer rises), a pressure sensor (activates dialer when boiler pressure of heating system exceeds preset limit), a glass break sensor, a flood sensor (alerts you when your premises are being flooded), an ultra-sensitive reverse pressure sensor (detects removal of valuable objects), or a hidden mat (alerts when an intruder steps on carpet or runner).

Facsimile Communications

Facsimile in engineering is the process of scanning graphical information, including pictures, converting the information into signal waves, and using these signals locally or remotely to produce a likeness, in record form, of the subject copy. A facsimile system consists of a facsimile transmitter, sometimes referred to as a scanner, transmission means, and a facsimile receiver. In some systems, such as the telephone, a converter is used to change the type of modulation to one more suited to the circuit used.

STRINGENT REQUIREMENTS

The nature of facsimile is such that it poses the most stringent of requirements on the overall system. Special consideration must be given to level change, minute impulse noise interference, low level sixty cycle pickup, small variations in transmission delay, echoes, multipath, envelope delay distortion, synchronization, and undesirable compression techniques. In the transmission of the million elements comprising a picture, one pulse received considerably different from its correct level, position, or sharpness becomes outstandingly apparent in the received picture, whereas with voice transmissions, considerably more distortion can be tolerated.

Facsimile is required to transmit a stationary image faithfully with a single frame, or *hard copy*. Unlike television, which consists of many frames, in facsimile, each image is an entity of its own.

EARLY FACSIMILE METHODS

In essential aspects, facsimile transmission and telegraphy are first cousins. Both are means of encoding information into electrical impulses and beaming the information from one location to another many miles away. Facsimile transmission and telegraphy were born at about the same time.

One of the earliest facsimile systems was patented by a Scot, Alexander Bain, in 1843. This was one year before Samuel F.B. Morse flashed the famous exclamation "What hath God wrought" between Washington and Chicago in the historic first long distance test of the infant telegraph.

Bain's facsimile system consisted of a picture made of shellac on a metal base explored by an electrified stylus mounted on a swinging pendulum. Impulses generated by contact of the stylus with the metal were transmitted to a receiving stylus moving across chemically treated paper. The paper would darken where the stylus touched and produced an image. However, systems based on mechanical movement of a stylus were limited in use to line drawings, written telegrams, and other simple transmissions. Development of telephotographic techniques greatly expanded facsimile capabilities.

Telephotographic systems were introduced around the turn of the century. A photo or other material was transmitted by means of the stylus or other instrument creating an electric current. At the receiver end, the current was converted into a narrow beam of light which swept across photosensitive paper. When this paper was developed, an image was created.

THE PHOTOELECTRIC CELL

Modern facsimile techniques emerged with the development of the photoelectric cell, a device which generates electric currents in response to light. In this system, a narrow, intense beam of light scans the picture and reflects light into a photoelectric cell. The intensity of the generated current is proportional to the amount of light reflected from each picture area.

The technological breakthrough which made facsimile transmission over ordinary telephone lines possible occurred around

1963. However, use was limited until the Carterfone case before the Supreme Court overruled objections by AT&T and established the right of the public to use telephone lines for non-telephone purposes.

Today, the use of facsimile communications is growing rapidly, not only in the industrial sector, but also in the public sector. Facsimile shipments are expected to reach $700 million by 1980. Anticipated reductions in telecommunications line costs will contribute to making facsimile communication even more widespread. The increasing availability of wideband communication channels, the use of satellites, and fiber optics cables for communications will all help to reduce line costs for facsimile transmission. Low cost equipment will also pave the way for increased facsimile equipment installation in your home.

These anticipated cost reductions, along with continuing refinement of facsimile technology, may provide the real push to the whole concept of electronic mail. Electronic mail may well be competitive with postal service charges, besides offering the tremendous convenience of getting and receiving mail in your home electronically in two minutes or less. The U.S. Postal Service itself is investigating the potential for facsimile as electronic mail carriers in its daily operations.

A major factor influencing compatibility between manufacturers equipment is the new CCITT (Consultative Committee for International Telephone and Telegraph) standards for facsimile communication. Recently de facto CCITT standards for high speed facsimile transmission have been agreed upon, and by 1980 they are expected to be made official. CCITT standards will regularize the specifications of page size, modem speed, compression scheme, and other technical features of facsimile communication. It is essential for companies to adhere to these standards in any new equipment they develop, or run the risk of equipment obsolescence.

Once the CCITT recommendations have been specified, the thrust of facsimile evaluation will shift from modems and speeds, which will have been standardized, to other considerations. Such factors as equipment modularity, the quality of the copies generated, and available interfaces may become the factors which separate one manufacturer's product from another and play a decisive role in determining which facsimile equipment you should select.

ELECTRONIC MAIL

Because of its speed, accuracy, and easy operation, facsimile equipment has become a major carrier of "electronic mail." It converts text, photographs, and graphics into electronic impulses which are transmitted over telephone lines. At the receiving end, the impulses are converted back into a facsimile of the original (Fig. 8-1). Placing your phone on a special transmission/receiving area and a rotating drum permits the transmission and receiving of this electronic mail. Facsimile reaches anywhere that telephone lines or satellite communications reach, and delays due to distance are eliminated. An urgent letter, recipe, or message can be sent across town or across the ocean with equal speed. All elements of the message, including signatures, are reproduced. Facsimile machines can be operated by anyone, and require no special training.

The unit shown is the dex^R 1100 facsimile system from Graphic Sciences, a subsidiary of Burroughs Corporation. This unit has a built-in "acoustic coupler" for transmission via standard phone lines.

Fig. 8-1. Facsimile unit dix^R 1100 by Graphics Sciences (Courtesy of Burroughs Corp.).

The device prints by means of a controlled voltage stylus operating on electro-sensitive paper. Output copy is provided in 8½ inch ×11 inch, or 8¼ inch × 11¾ inch size paper. Scanning is accomplished by a mobile head moving along a rotating drum. Input documents may range up to 8½ inch × 11¾ inch.

Vertical resolutions may be selected to fit requirements for communications speed and copy quality. The unit is designed to operate at ninety-six lines per inch at six minutes, and sixty-four lines per inch at four minutes in the FM mode. Variations of the unit, models 1102 and 1103, operate with different resolutions and speed.

Of course, for facsimile operations, you need one unit at your end and one at the other party's end. In order for facsimile scanners to interoperate with the receiver on the other end, certain characteristics must be considered. The scanning line frequencies (rpm) of the scanner and recorder must be essentially equal. The tolerance is usually between one part in 150,000 and one part in 300,000. The index of operation ensures the proper ratio between copy width and copy length.

The direction of scanning is the manner in which rectilinear scanning takes place. A normal scanning direction implies scanning in the same manner as a person reading, that is, the first scanning line starts from the upper left corner of the page and proceeds to the right. Subsequent lines follow from top to bottom. A wrong direction of scanning results in a mirror image and is sometimes purposely used where a negative recording is desired.

Three techniques are prevalent for synchronization: tuning fork or crystal control of scanning at each end of the circuit; transmission of a reference synchronizing signal (sometimes a 60 cps modulation of a carrier); and the control of the scanning rate by a common commercial power line frequency. The synchronization of the scanner and recorder must be compatible.

The type of modulation, such as AM or FM, and the direction of the modulation must be compatible. The contrast going from white to black, and the linearity should also be matched. The transmitter drum factor must not exceed the receiver drum factor.

SCRAMBLING FACSIMILE COMMUNICATIONS

Of course, facsimile transmissions are just as susceptible to interception as are voice transmissions. It's no wonder, then, that with the increased use of facsimile equipment, there's an increase in the availability of scramblers.

Technical Communications Corporation has developed several units for scrambling facsimile communications. One such device is their FX-703 Multicode Facsimile Privacy System, designed for FAX (facsimile) transmissions over FM, AM, SSB and telephone communication systems having bandwidths as narrow as 2.3 kHz (300 Hz to 2300 Hz). Interface parameters, such as levels, impedances, and connections are easily changed by use of internal switches and programming plugs.

The unit utilizes an analog audio scrambling technique operating in the time and frequency domains, controlled by digital codes and synchronized by an FSK code burst which is a function of the code set into the unit. These can be varied by the front panel code selector switches and the use of strapping options located on readily changeable jumper plugs, thereby providing a wide range of code selections plus a high degree of built-in code protection. Thus, the simple combination of front panel code selection switches and strapping options provides control of three independent codes, namely synchronization, timing and frequency. By this means, the scrambling code repetition cycle can be extended to approximately 1200 hours.

In this unit's scrambling process, the audio signal is switched by band pattern frequency shifting and time interval changes under control of digital codes. Eight frequency increments are used, with both time and frequency intervals varying between steps.

The combination of front panel switching and internal strapping options provide a total of 10^{16} useable codes, and the front panel switches alone permit any one of 262,144 codes at the time to be selected. Simple internal changes can then be made to provide other blocks of code groups as required. The level of security of this device cannot be defeated by conventional decoding techniques, and it is also highly resistant to sophisticated electronic deciphering methods.

Mobile Radiotelephones

Since a telephone operates on the principle of "communication by wire," and an automobile is a moving vehicle, it is obvious that the initial connection between the vehicle and the telephone network has to be through the airwaves. Therefore, a mobile telephone (Fig. 9-1) differs from a regular phone mainly in that radio is the transmission medium to the central office instead of a land line.

Mobile phones have been around for about thirty years. In those days, you had a selective signaling encoder in your car which rang a bell when you were called. Outgoing calls were made by giving the number you wanted to call to the Mobile Service Operator. Ten years later the system of direct mobile dialing was developed (Fig. 9-2).

There are two basic mobile telephone systems: MTS, Mobile Telephone Service, mobile units that are operable on one or more radio channels, with manual channel selection; and IMTS, Improved Mobile Telephone Service, making use of multi-channel mobile units with automatic channel scanning and outgoing dial capability, rotary or Touch-Tone.

Presently being implemented on a trial basis is the Cell System developed by the Bell Telephone System. While the other systems operate in the 150 MHz and 450 Mhz bands, the Cell System operates in the 900 MHz region. The concept, theoretically, is simple. In a conventional mobile phone system, one high powered

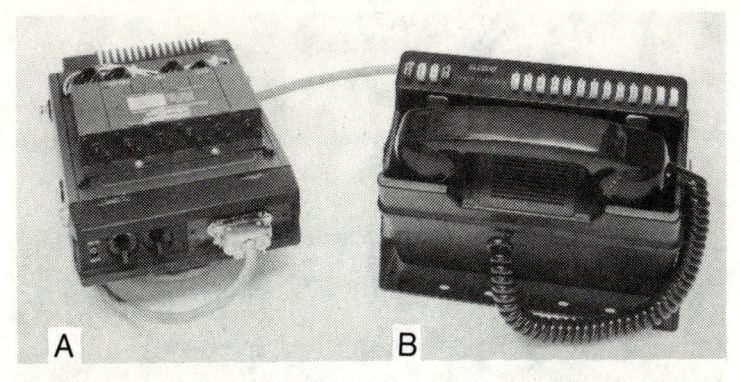

Fig. 9-1. Mobile telephone component: A, transceiver; B, control head (Courtesy of Standard Communications Corp.).

central transmitter and receiver are serving all mobile units in the area, thus allowing a single frequency to be used by only one mobile unit at a time. In the Cell System, many smaller transmitters/receivers, with each covering only a few square miles, are installed. This way, a given frequency can be used simultaneously by mobile units in several different areas, or "cells" without interfering with each other. The result: a lot more calls can be placed on a given frequency band.

While the system is simple in principle, it is enormously complicated in practice. How is it decided which mobile unit gets which frequency at which time? How can it be ensured that two adjacent cells are not using the same frequency simultaneously and causing interference? The answer is a complicated system of computer control, which we will not discuss within the framework of this book. Besides, it is presently only an experiment.

If you want phone service in your car, you can buy the equipment or lease it from the Radio Common Carrier that furnishes mobile telephone service in your area. In addition to mobile telephones that are installed in a vehicle, there are portable mobile telephones packaged in an attache case. Besides the cost of buying or leasing the equipment, you have to pay a minimum monthly service charge for mobile telephone service.

When you do get phone service, you do not get a private line (radio channel), but you share the circuit with up to seventy other subscribers. If you have a multi-channel phone, your chances of

Fig. 9-2. A mobile telephone in use (Courtesy of Standard Communications Corp.).

finding an unused circuit are enhanced. Since your mobile phone is equipped with a selective signaling decoder, you just wait for the phone to ring and then answer the call.

Mobile telephone service is furnished by regular telephone companies and by RCC, Radio Common Carrier, companies. If you lease the unit from the Common Carrier, your phone is covered by that station's license. If you furnish your own unit, however, you have to get your own mobile station licence from the FCC.

The beauty and convenience of a mobile phone is that you take your telephone with you wherever you go, giving you access to the phone service, almost as if you were at home.

HOW MOBILE TELEPHONES WORK

Basically, your mobile phone consists of a transceiver (to connect you over the air with the phone company or RCC), a control head (for remote control of the transceiver), and a phone patch.

Since a transceiver operates on the "simplex" principle (push mike button to talk, release to listen), and the telephone on the "duplex" principle, a phone patch is an electronic device which interconnects the duplex telephone audio circuit with the simplex radio circuit. The VOX, voice actuated circuit, in which a relay provides transmitter keying when the phone party talks, relies upon voice traffic to perform its function. The first design parameter for the phone patch, therefore, is the voice audio frequency spectrum. The typical two-way radio audio bandpass is 300 Hz to 3000 Hz, and as a result most phone patches are designed to operate within that same frequency spectrum.

Another variable is the telephone circuit level. Assuming a clean phone circuit, long distance connections normally provide a lower audio signal than local connections. Also, people talk at different levels. This problem is solved by using a compression amplifier to provide a constant audio level to the transmitter, and is adjusted so that the level it provides is comparable to the base station operator talking over the microphone at a normal voice level.

The best way to describe a mobile telephone system is to follow a description of the operation of an actual system, in this case, one by Astronautics Corp. of America.

The unit (Fig. 9-3) is a solid state, UHF frequency modulated (FM), IMTS Radio Telephone designed for use with the Bell System or Radio Common Carrier Automated Mobile Telephone Systems.

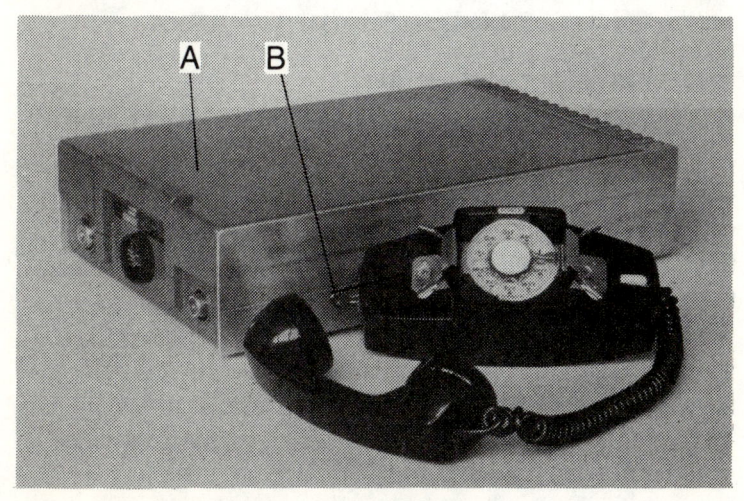

Fig. 9-3. Mobile telephone: A, receiver/transmitter; B, remote control head.

The unit consists of a receiver/transmitter (Fig. 9-3A), remote control head (Fig. 9-3B), an antenna, and appropriate interconnecting cables and connectors.

The receiver/transmitter (FCC Type Acceptance # R400-T400) contains the receiver, transmitter, digital supervisory unit, duplexer and power supply circuitry. The unit is contained in a rugged case suitable for mounting under a seat or in a trunk. Two front panel connectors provide provisions for 12V DC input power, connections to the remote control head, and a connection to the antenna. Frequency coverage is 450 MHz to 470 MHz, with a 5 MHz difference between receive and transmit frequencies.

The integral cover mounting arrangement secured with a key lock is designed to minimize tampering or unauthorized removal of the radio, once installed. A handle is provided on the front panel of the receiver/transmitter to aid in removal and to facilitate handling.

The control unit (Fig. 9-3B) consists of a dial, telephone handset, and the various controls and indicators required in the operation of the radio. These include: on-off switch, home or rove channel selector, horn selector, power-on indicator, transmit indicator, and busy indicator.

The antenna is designed for rooftop installation, but can be mounted in any convenient location. The antenna provides 5 dB gain with the required 50 Ω load to the transmitter at less than 1.3:1 VSWR. See Table 9-1 for specifications.

STATION TO MOBILE UNIT CALLING

The unit, when turned on, scans each channel in its complement (twelve channels) until the idle tone transmitted by the base station (the phone company or RCC) is detected. The presence of idle tone on any one channel causes the supervisory unit to latch the radio to that marked idle channel. When a call for a mobile unit is received at the telephone control terminal (phone company or RCC), the base station indicates the marked channel "busy" by replacing the idle tone with a seize tone. Upon receipt of the seize tone, the supervisory unit inhibits the receiver/transmitter from originating a call and prepares to receive the coded calling signals from the base station.

Shortly after replacing the idle tone with a seize tone, the base station transmits the calling code (telephone number of your mobile

Table 9-1. General Specifications of the
Mobile Phone by Astronautics Corp. of America.

General Specifications:	
Size	
Receiver/transmitter	3⅞" × 17⅞" × 10"
Control unit	6¼" × 2½" × 5½"
Weight	
Receiver/transmitter	25 lbs.
Control unit	1 lb.
Primary voltage	12V DC, negative ground
Primary current drain	receive: 0.5 amps
	transmit: 6.0 amps
Operating temperature	−30° to +55°C
Transmitter Specifications:	
Frequency	450 MHz to 470 MHz
Multichannel operation	12 channels
Frequency stability	± 0.0005%
RF Power	20W nominal
Spurious & harmonics	spurious emissions >70 dB
	harmonics >60 dB
Modulation	limited to ± 5 kHz
Audio sensitivity	0.18 VRMS at 1000 Hz − 3.33 kHz deviation
Receiver Specifications:	
Selectivity	>70 dB at ± 17.5 kHz
Sensitivity	0.6 μV for 12 dB sinad

phone) of the desired mobile. Each digit of the code is comprised of alternate fifty millisecond bursts of idle (2000 Hz) and seize tone (1800 Hz), interrupted by approximately 300 milliseconds of seize tone between each digit. The number of idle tone bursts corresponds to the digit transmitted. For example, if the code digit to be transmitted is three, then three bursts of idle tone are sent in the code pulse train (Fig. 9-4). These pulses are counted by each mobile unit and compared with a code preset into the supervisory unit. If this unit correctly receives all seven digits of the calling code, it turns on the transmitter and transmits an acknowledge signal (750 milliseconds at 2150 Hz). Upon receipt of the acknowledge signal, the base station sends a series of alternate idle and seize tone pulses to actuate the control unit buzzer. When you remove the handset from its holder (off-hook) to answer the incoming call, the supervisory unit inhibits the search for idle tone, turns on the transmitter and sends 400 milliseconds of connect tone (1633 Hz). While the transmission is in progress, the transmit lamp is turned on in the control unit. After the transmission of connect tone, the handset is activated to allow conversation to follow.

When your call is completed and the handset returned to its holder (on-hook), the supervisory unit sends a 750 millisecond disconnect signal of alternate twenty-five millisecond pulses of disconnect (1633 Hz) and guard tones (2150 Hz). After transmission of the disconnect signal, the supervisory unit turns off the transmitter and proceeds to scan through the channels looking for a new marked idle channel. If a wrong number is received for any of the digits in the calling code (area code and station number), and therefore a mismatch has occurred, the supervisory unit releases from the latched channel and commences to search for a new marked idle channel.

MOBILE UNIT TO BASE STATION CALLING

When you lift the handset of your control unit (off-hook) to initiate a call, and the supervisory unit is receiving idle tone at the time of off-hook, this unit inhibits itself from hunting for a marked idle channel, turns the transmitter on, and transmits guard tone for 350 milliseconds (Fig. 9-5).

If, after receiving the off-hook signal from the control unit, the supervisory unit fails to receive idle tone for a period of 350 milliseconds (channel seized by another mobile unit) or continues to receive idle tone for more than 400 milliseconds (transmission of connect tone not received by base station), the supervisory unit blocks the call, turns off the transmitter, and turns on the busy light on the control unit. After the 350 milliseconds of guard tone, the supervisory unit transmits, in normal operation, fifty milliseconds of connect tone.

Upon receipt of the connect tone from your mobile phone, the base station quickly removes the idle tone from the channel. After a pause of at least 250 milliseconds, the base station transmits a minimum of fifty milliseconds of seize tone and gets ready to start receiving your mobile's identification number (area code and station number). When the base station removes the seize tone, the supervisory unit starts generating 25 millisecond pulses. For every odd number pulse generated (1st, 3rd, 5th, etc.), the supervisory unit sends twenty-five milliseconds of connect tone, and then 25 milliseconds of unmodulated carrier. For the even numbered pulses

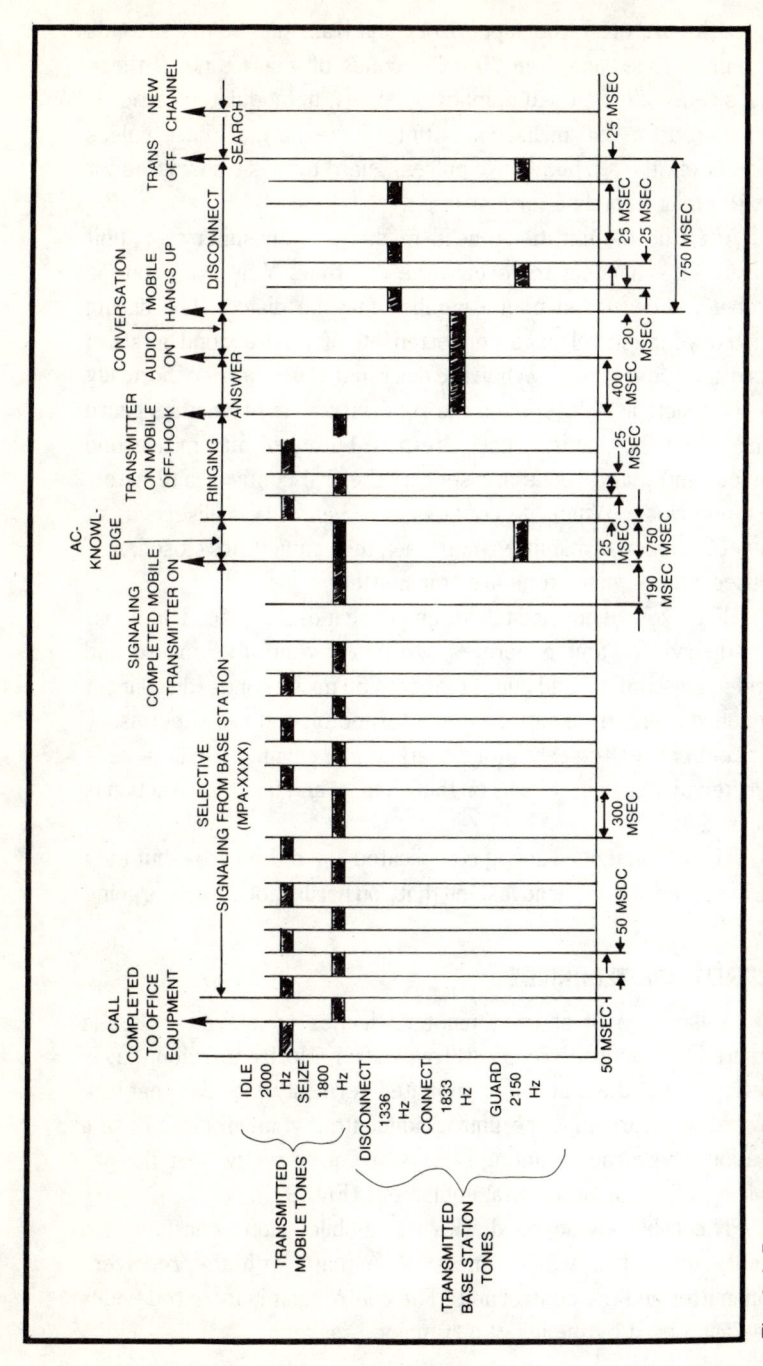

Fig. 9-4. Base-to-mobile call signaling tones sequence chart.

(2nd, 4th, 6th, etc.), the supervisory unit transmits 25 milliseconds of connect tone, and then 25 milliseconds of guard tone. If these pulses cease after an odd number of pulses, unmodulated carrier is transmitted for 190 milliseconds (interdigit time); if these pulses cease after an even number of pulses, guard tone is transmitted for the 190 milliseconds interdigit time.

After the identification tone is transmitted, the supervisory unit activates the handset to receive the dial tone. When this tone is received, the desired phone number may be dialed. The dialing action of your control unit is converted into fifty millisecond bursts of guard and connect tone. When the dial is not in use, a set of normally open contacts are closed, and the supervisory unit transmits guard tone. When the dial is in use, alternate bursts of fifty millisecond connect and guard tones are sent as the dial contacts alternately open and close. When the contacts are open, fifty milliseconds of connect tone are transmitted, and when the contacts are closed, fifty milliseconds of guard tone are transmitted.

When you've finished talking and the handset is placed on-hook, the supervisory unit generates alternate twenty-five millisecond bursts of disconnect and guard tone for 750 milliseconds (disconnect sequence). The transmitter is then turned off and the supervisory unit commences to search for a marked idle channel. If idle tone is detected by this unit, it locks on that channel and the search action is stopped.

These sequences sound complicated—and they are—but they are executed in such rapid fashion that you hardly notice what's going on.

INSTALLATION TECHNIQUE

A flat area of at least nineteen inches by eleven inches is required to mount the receiver/transmitter. Horizontal mounting is preferred, but the unit can be mounted vertically with the heat sink upward. Be sure that the unit is adequately ventilated. Choose a location where the mounting screws are not directly over the gas tank, gas line, or other vital equipment (Fig. 9-6).

The cable kit supplied with the mobile phone consists of a twenty foot cable with connectors to mate with the receiver/transmitter and the control unit. The control unit is mounted under the dash board by means of a trunnion bracket.

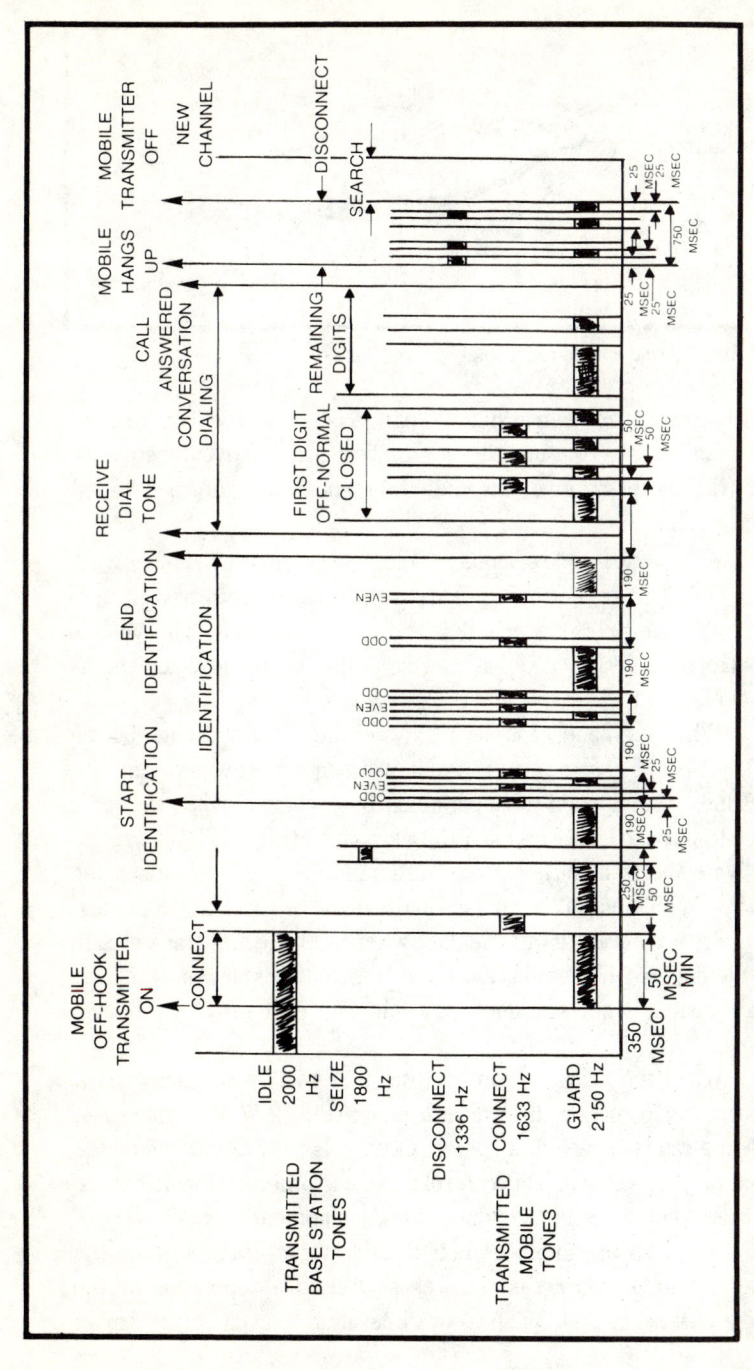

Fig. 9-5. Mobile-to-base call signaling tones sequence chart.

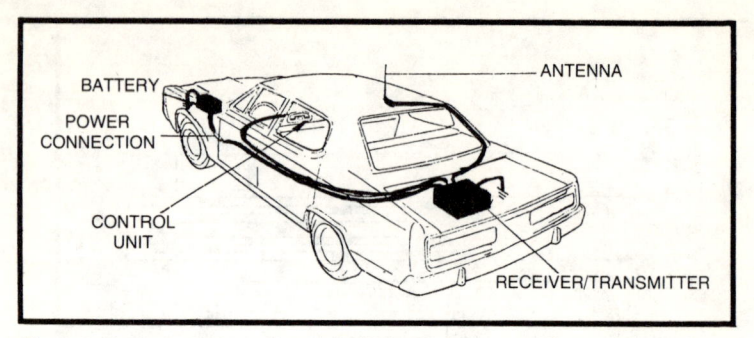

Fig. 9-6. Typical mobile radiotelephone installation.

After you've been assigned a phone number, you have to code this number into your unit. To accomplish this, remove the cover of the transmitter-receiver and code the number with the color coded digit wires.

Another example of a mobile telephone control head is shown in Fig. 9-7. The key lock (Fig. 9-7A) turns power on to transmitter/receiver and prevents unauthorized use of the unit. The volume control (Fig. 9-7B) located at the rear of the control head adjusts the audio level in the earpiece.

When the handset is off-hook, and the channel change button (Fig. 9-7C) is depressed, the unit will sequentially go to another channel. This is used to select a non-busy channel when making a mobile-to-base call. The push-to-talk control (Fig. 9-7D) is used to activate the transmitter in the MTS mode of operation, or in the party-IMTS mode. It is not used in the home or roam-IMTS modes. It is also not necessary to use this switch after the first transmission in the duplex-MTS mode, since the transmitter stays on when the DUP switch is on. The duplex switch (Fig. 9-7E) is only for MTS mode.

When the auxiliary control (Fig. 9-7F) is on, a flashing AUX indicator appears on the indicator panel (Fig. 9-7G). In this case, when a call is received, a contact closure is provided to sound the horn on your vehicle. This control panel has other light indicators to indicate the status of the controls and equipment.

When calling another mobile unit in the IMTS mode, the operator in certain areas requires both mobiles operating on the same channel to go to push-to-talk operation to avoid interference.

Flip this switch (Fig. 9-7H) only when directed to by the mobile operator. When this switch is on, the DUP light will flash. Control is automatically turned off when the handset goes back on-hook.

The operation mode switch (Fig. 9-7J) is a rotary three-position switch. Set in *home* position for IMTS service in your local area. Switch to *roam* for IMTS service outside local area. Switch to *man* for manual MTS operation where this type of service is offered.

The channel selection switches are eleven miniature switches at the front bottom of the control head. These are generally set by your service center in your home area.

As with the other telephone operations described, mobile telephone can also be scrambled for security reasons. Communications Security Devices offers a unit for such voice privacy.

Using a combination of frequency inversion, tone masking, and sideband masking, scrambling is accomplished. Their Series 107 has

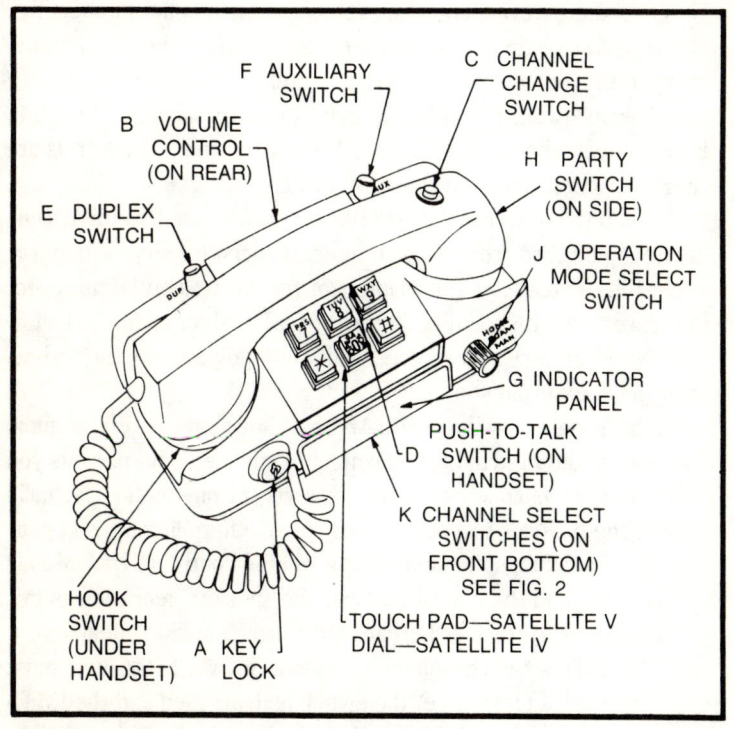

Fig. 9-7. Push-button mobile control head (Courtesy of Livermore Data Systems).

twenty-five codes in each of four code families for a total of 100 combinations in the system. Codes are changed by a plug-in module. Both telephone and radio versions are available and can be supplied so one may communicate over a link consisting of a radio channel and a telephone channel in series.

ATTACHE PHONE

Should you leave your car and not want to be without telephone service, Livermore Data Systems also supplies a briefcase-type telephone, the Attache Phone (Fig. 9-8). This Attache Phone is capable of operating through either the operator accessed Manual Telephone System, MTS, or the two-way dial service, IMTS. All eleven channels of the VHF Mobile Telephone Service are provided. The transmitter delivers eighteen watts of output power to either the internal antenna, concealed behind the pocket in the attache case lid, or to the external antenna. Your personal phone number, five digits for MTS service, or seven digits for IMTS service, is programmed by your dealer or service center.

The following controls and indicators are shown in Fig. 9-8. Except for the hookswitch and push-to-talk switch, all controls are alternate action, push-on, push-off switches.

When the power switch (POWER) is depressed, your Attache Phone is powered from its own internal Nicad battery. Whenever either the cigarette lighter adapter, or the AC wall outlet are used, the power switch should be in the *off* (up) position. This will allow your Nicad battery to be recharged while power is being derived from these external sources.

The speaker switch (SPEAKER) allows the received audio signals to be directed to the audio monitor speaker. This permits you to monitor the channels to listen for a vacant one without actually holding the handset to your ear, or to let others listen in on your conversation when so desired. If the volume control (VOLUME) is set too high, and the handset is near the speaker, feedback as in a public address system will create a disturbing audio squeal.

The IMTS switch selects the mode in which the unit must operate. For IMTS service, the switch is depressed and the IMTS indicator will light. For MTS service, the switch should be *off* (up), and the IMTS indicator unlit.

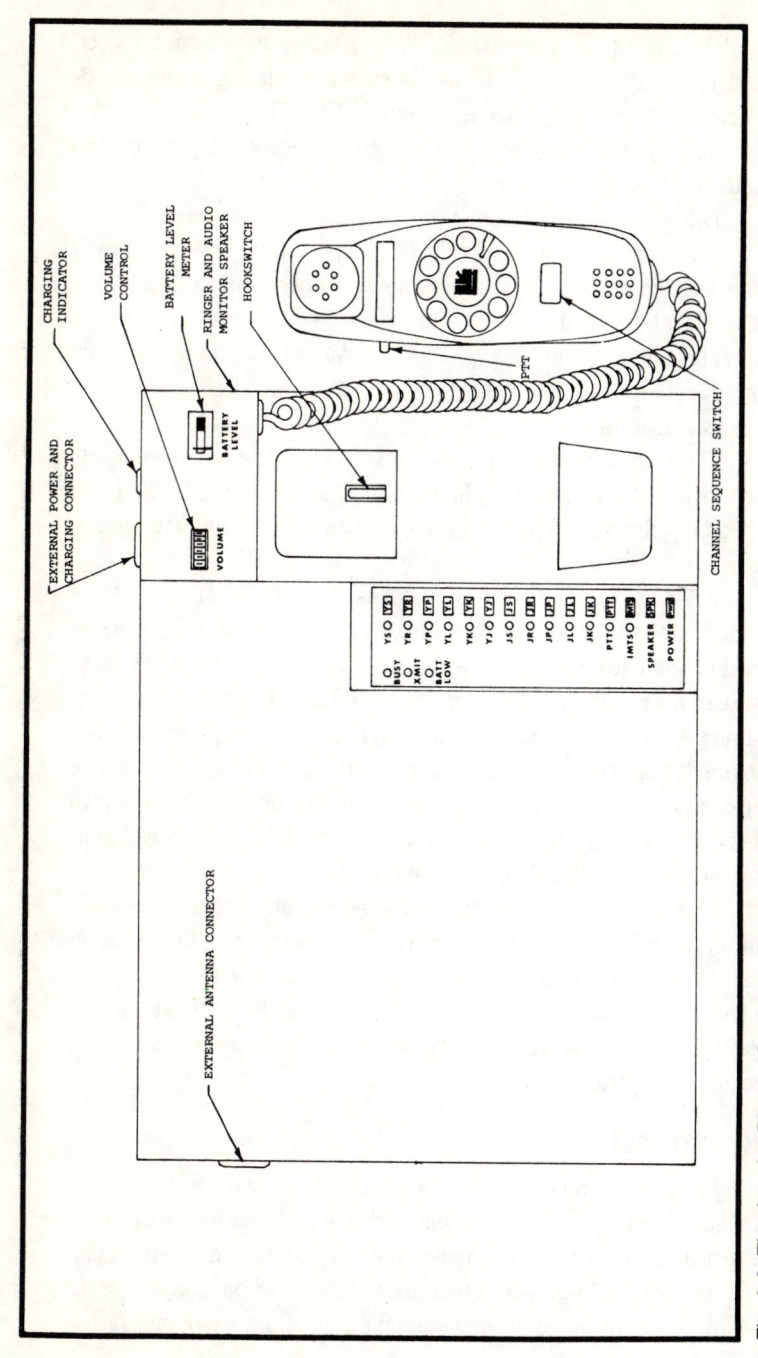

Fig. 9-8. The Attache Phone (Courtesy of Livermore Data Systems).

The push-to-talk switch (PTT) is a momentary action switch used primarily on the IMTS network when placing a mobile-to-mobile call over a single IMTS channel. The PTT indicator can be turned off by alternately pressing the button, or by "hanging up" the handset on hookswitch.

The eleven two letter switches represent the eleven VHF two-way channels that are available for use throughout the North American continent. Attache Phone will only scan/search those channels whose switch buttons are depressed.

The battery indicator (BATT LOW) warns you when your battery level is low, so you should recharge the battery when this indicator lights up.

The transmit indicator (XMIT) is lighted whenever your Attache Phone is radiating eighteen watts of power. (The battery, when fully charged, permits forty-five minutes of continuous transmission).

When the handset is off-hook, a busy lamp (BUSY) on the MTS mode shows the channel is busy. In this case, to make a call, choose another suitable channel. In the IMTS mode, a BUSY lamp ON, with handset off-hook, indicates that you have failed to seize the channel for one or more reasons: the channel is busy, the equipment was not locked on the idle tone, or the range to station is too great. In any of these cases, you must return to on-hook position before trying to make another call. In on-hook position, the BUSY lamp will light when the unit is decoding an incoming call.

Operation of the Attache Phone is like any other telephone, within the parameters of the controls just described. The Attache case lid must be in a vertical position to fully utilize the built-in antenna. Keep metal objects away, and orient the handset cord so that it is horizontal and away from the antenna to prevent undesirable pickup of the transmitted signal.

THE SKYPHONE

Flying in a private plane, and you want to maintain telephone contact? Well, there's the Skyphone[R] from Astronautics Corp. of America which allows you to make a call while you're flying from one point to another (Fig. 9-9). Operating in the UHF band of 450 MHz 470 MHz, the plane-to-ground station range of the Skyphone varies

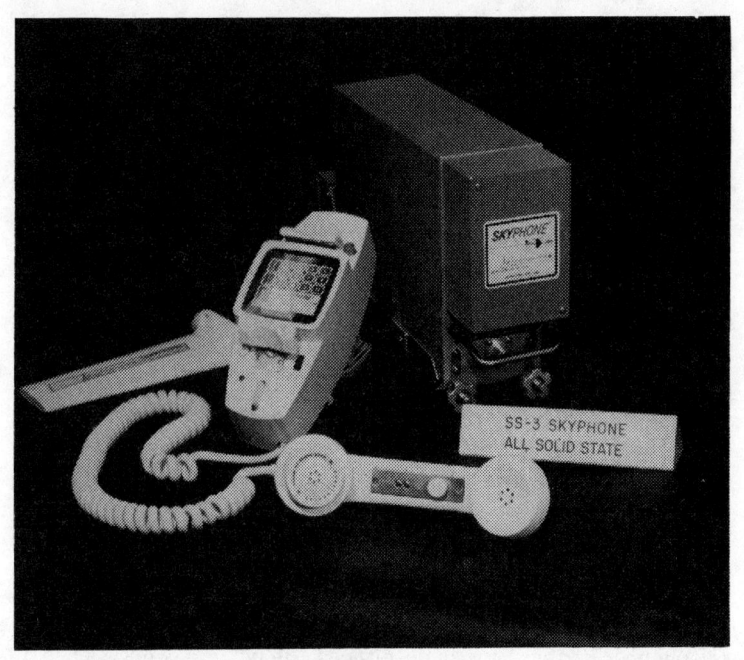

Fig. 9-9. The Skyphone[R] (Courtesy of Astronautics Corp. of America).

with the altitude of the aircraft. At 10,000 feet, the range is 175 miles; at 20,000 feet, 275 miles.

As a multi-channel device, the Skyphone gives you a sufficient number of two-frequency talking channels to provide nationwide airborne telephone service. This feature has allowed the development of an interference-free ground station system that provides telephone capability over major air routes at 5000 feet and up, as well as multiple-channel coverage in some high-density areas.

The Skyphone will not interfere with two-way radio equipment aboard the plane. A call is placed as you would any long distance phone call with your home phone.

GOVERNMENT REGULATIONS

Each user of mobile radio equipment is subject to the provisions of the Communications Act and the rules and regulations of the Federal Communications Commission. Persons violating these rules may be denied service and are punishable by law.

State your mobile phone number at the start and end of each call. No person may knowingly transmit a false or fraudulent distress signal, and obscene, indecent, or profane language is prohibited.

The willful interference with any call or signal is illegal. It is unlawful to "listen in" on any conversation intended for others, or to divulge any information so obtained. All instructions given by the operator must be followed since the mobile unit is under the control of the base station licensee.

A check for satisfactory equipment operation must be performed annually, and only FCC licensed technicians may repair or adjust the equipment.

Finally, you are responsible for all calls made from your equipment. Lock your unit when it is to be left anywhere where unauthorized access is possible.

THE RADIO COMMON CARRIER

A Radio Common Carrier, RCC, obtains a license under Part 21, Subpart G of the Communications Rules. He may offer mobile telephone, dispatch, or paging services. He generally must obtain a certificate of convenience from a state or local Public Utilities Commission. It is before the PUC that he submits his proposed tariffs, or the charges to the subscriber for the service.

To obtain a license, he must show that he is legally and financially qualified, that there is a need for common carrier service in the area he proposes to service, and that the rendition of such service will neither technically interfere with or cause economic injury to another common carrier.

The license he obtains gives him exclusive rights to the channel(s) assigned to him in his area, and he can protest any other common carrier application on the grounds that he will suffer interference or economic injury. Because of his status as a common carrier, he can adjust his charges to assure whatever guaranteed rate of return the state or local PUC has determined to be reasonable.

By definition, a common carrier offers services for hire to anyone and everyone. His customers are, therefore, private citizens and also some who could obtain their own license in one of the private land mobile services, such as the Business Radio Service. The

RCC's customers do not obtain their own licenses, but rather are covered by the RCC's authorization.

The RCC has total control of his channel(s). All transmissions go through his base station. He makes the radio connection either by his operator, or automatically, thus enabling the call to be completed. He can thereby terminate or cancel your service, or refuse to handle any subscriber's call. He alone is responsible for the proper operation and maintenance of the system, and he has the sole right of access to the property on which his station is located.

COMPUTER TECHNOLOGY AND THE MOBILE PHONE

With the advent of computer technology, the latest mobile telephones are using microprocessors to perform a number of tasks, such as monitoring all operator-controlled switches, controlling the transmitter/receiver, comparing a received code (idle tone, or connect tone) with the specific code in the control unit, and controlling various status indicators to the operator.

Motorola's Pulsar II control unit utilizes microprocessors to allow you to have abbreviated dialing of up to ten numbers. These ten numbers are selected and programmed by you from the dial pad and can be easily changed. The underside of the handset contains a handy reference directory to identify the stored numbers and their memory location digit.

The unit features pushbutton dialing, allowing you maximum ease and speed of dialing. The pushbutton pad is located on the back of the handset so you can dial with one hand.

On-hook dialing allows a phone number to be entered from the dial pad with the handset in the cradle. You simply obtain the dial tone and push the SND button to initiate the call. A LED channel number display identifies the particular channel being sequenced in the manual mode, reviews the channels programmed in the *roam* or *home* mode of operation.

The key pad and control panel are softly lighted, and an electronic ringer provides a pleasant call alert. If you remove the handset or push the off-hook button and the channels are busy, the busy light will provide visual indication, along with a busy tone in the handset.

The control head can be programmed to scan the desired channels while on-hook. In the HOME mode, the unit will scan only those channels available in the home city. These channels are programmed in the unit and prevent it from locking in on a foreign, unwanted channel.

The key to the microprocessor's success is its ability to "remember" the jobs it must perform. The microprocessor, abbreviated MP, processes information in rapid, sequential manipulations at the rate of hundreds of thousands of bytes (a group of adjacent binary digits often shorter than a word that a computer processes as a unit) of information per second. The microprocessor has endeared itself to hundreds of engineers because it can be programmed to perform a variety of dedicated functions faster, easier, and in some cases more reliably than a combination of bulkier conventional integrated circuits. In addition, many microprocessors can be "reprogrammed" if necessary to perform tasks other than those for which they were originally programmed, thus minimizing obsolescence. The microprocessor is thus smaller in size, has higher component density (important in mobile application), greater flexibility and lower maintenance costs for the finished product.

Computerized hardware for on-the-go telephone calls is readily available, but the radio frequencies needed to carry the heavy volume of mobile calls are simply not available. Because radio channels are so limited, there are long waiting lists in most cities for mobile phone service. To cure this congestion, a consortium of three RCCs has filed an application to try the technique of using digitalized voice signals. This RCC technique, developed on paper by Harris Corp., requires one extremely powerful (375 kW) transmitter site. Voice signals would be digitalized and beamed out as bursts of pulses in packets, each packet coded for separate mobiles.

Another concept, known as *spread centrum*, is receiving attention among those who want to solve the congestion problem of the mobile telephone. Used extensively by the military, spread centrum signals are highly immune to jamming and interception. Imagine that each FM station spread its signal across the entire FM band from 88 MHz to 108 MHz. Each station, however, would encode its output so that a special filter in your FM set could decode that signal. For mobile telephone communications, recently developed semiconductor and electronic-filter technologies might make it possible for everyone in the country to have a unique spread centrum decoding circuit for a traveling phone.

The Future

A boundless future is seen for the telephone. Versatility will pervade network service. We will have computerized automated repair service bureaus, toll-service positions, and directory retrieval systems that have more and more operators sitting at CRT consoles (TV-screen type consoles) when they answer a call. A directory-assistance operator taps four letters of a last name, hits a "book" button to choose which city or section to search and, in less than two seconds, gets both a microfilm display of one of 72,000 printed pages on her CRT display.

Bell Telephone Magazine predicts a take-along phone in the not too distant future. Others are talking about the wristwatch telephone, weighing three ounces and costing about $10, and transmitting on only twenty-five milliwatts of power. Yet, the wristwatch phone will instantly get you in touch with 2.5 million other subscribers on almost any telephone in the world, with up to 100 people on a party conference line at once.

High-powered seven kilowatt satellites with large antennas 200 ft. in diameter relay your signals from space. Your wristwatch phone also functions as a personal navigation system, showing you your exact position on the earth's surface. It is a link with a computer that ascertains your location. This phone can also be used as a private emergency hot line, connecting you instantly with police or medical services for help.

There are already a number of communication satellites in space. More will follow, large, sophisticated satellites, each capable of providing more than 30,000 voice or data circuits simultaneously. They will operate in super-high frequencies that have not been considered feasible for commercial use until recently.

The intermediate step to the wristwatch phone will be an advanced Touch-Tone[R] telephone that will be capable of arranging financial transactions via a computer with your bank or any other institution. It will even be able to lock your front or back door, or to turn your oven off with the touch of a few buttons on your car phone.

These future Touch-Tone[R] phones will be capable of responding to the human voice via an interconnect with a unit that will recognize the sound of the human voice by digital means, and convert these voice sounds into required tones. In this way, you can control computers, telephone switching systems, and other devices, by simply speaking to them. Since your voice is as much a part of you as your handwriting or fingerprint, you can handle all sorts of personal business with your voiceprint.

In these days of actual energy shortage, we may find it more and more necessary to use electronic communications for our personal contacts, such as sending letters, checks, etc.. Yet, today, there is still an enormous gap between the telephone and mail. Large corporations have been able to bridge this gap with Telex, TWX, private-line teletype, private-line facsimile and private-line data transmissions. Although private users, such as you and I, have access to these communications modes, high costs have prevented extensive use of these devices by small companies or private individuals. New technologies such as component miniaturization will soon allow many more people to own these non-voice units.

Telephone companies are working hard toward the realization of a "communications society." A Stored-Program Controlled (SPC) network carries information on a separate high-speed channel in the form of electrical pulses. Bell Telephone System's Magnetic Bubble Memory stores information using tiny magnetic bubbles which are tiny magnetized areas in a thin film of material such as garnet crystal. The bubbles can be moved about to store and access data. Bell's Mac-8 Microprocessor is a device less than one-tenth the size of a postage stamp, but as powerful as the central processing unit of a small computer, capable of executing over a hundred thousand electronic logic functions per second while using only one-tenth of a watt of power.

What do telephone companies expect to happen during the next two decades? They expect greatly increased use of the mobile telephones, and dramatic cost decreases in long-distance and international phone calls. They expect increased use of satellites for data, facsimile, and TV communications, with increased use of data banks and data retrieval by telephone.

Telephone companies expect extensive telecommunications use in medicine. "Prediagnosis" interviews can be carried out between a patient and a distant computer, often to determine whether or not the patient should see a doctor or visit a hospital. Computers can automatically monitor chronically ill patients. Readings of their condition can be recorded by a tiny device that is linked to the telephone, transmitting data to the hospital's computer.

Telephone companies expect Picturephone[R] use will spread rapidly, with computer voice input systems that permit you to speak to a computer over the telephone, using a very limited vocabulary of clearly separated words. This type of telephone service could include telephone "dial-up" of various services such as TV programs, news services, flight information, and shopping areas. A directory the size of the "yellow pages" lists all the telecommunication services available and gives the codes necessary for communicating.

This future world we will live in demands and depends on skill in communication and in knowledge relevant to communication to an extent far beyond anything previously known. In the long term, satellites used in the telecommunications network will provide a nervous system for mankind, knitting the members of our species into a global society.

The technology for a world communications network is here. The challenge to us as individuals is to develop a public philosophy capable of supporting the critical choices which must be made to govern this brave new world.

Appendix I
Glossary

abbreviated dialing—Ability of your phone system to dial only two to four digits, while the network dials the balance of the seven to fourteen digits required.

address—A coded representation of the destination of computer data, or their originating terminal.

alternate trunk routing—The ability of the switching equipment to select up to five alternate routes if the route first accessed by the user is busy.

amplitude modulation—One of three ways of modifying a sine wave signal in order to make it "carry" information. The sine wave or carrier has its amplitude modified in accordance with the information to be transmitted.

analog data—Data in the form of continuously variable physical quantities.

analog transmission—Transmission of a continuously variable signal, as opposed to a discretely variable signal. The normal way of transmitting a telephone or voice signal has been analog. Now digital encoding, using PCM (see *Pulse Code Modulation*) is coming into use over trunks.

attendant busy verification—Permits attendant to verify whether the station is busy, idle, or out of order.

attendant hold—Incoming central office calls are held at the switchboard until the operator can process them.

attended operation—In data set applications, individuals are required at both stations to establish the connection and transfer the data sets from voice mode to data mode.

attenuation—Decrease in magnitude of current, voltage, or power of a signal in transmission between points, usually expressed in *decibels*.

audio frequencies—Frequencies that can be heard by the human ear, usually between 30 to 20,000 cycles per second.

automatic calling unit—A dialing device which permits a telephone to automatically dial calls over the phone network.

automatic dialing unit—A device capable of automatically generating dialing digits.

bandsplitting—A form of audio scrambling where the audio channel is divided into two or more frequency sub-bands and transposed into frequency.

bandwidth—The range of frequencies available for signaling. The difference expressed in cycles per second (hertz) between the highest and lowest frequencies of a band.

baud—Unit of signaling speed. The speed in bauds is the number of discrete conditions or signal events per second. If each signal event represents only one bit condition, baud is the same as bits per second, but if each signal event represents other than one bit, baud does not equal bits per second.

bel—Ten decibels (see *decibels*).

bit—Contraction of "binary digit," the smallest unit of information in a binary system. A bit represents the choice between a mark or space, one or zero, condition.

bit rate—The speed at which bits are transmitted, usually expressed in *bits per second*.

breaking—The act of decoding or extracting intelligence from a scrambled signal other than by using a matching scrambler.

broadband—Communications channel having a bandwidth greater than a voice grade channel, and therefore capable of higher speed data transmissions.

Broadband Exchange—Public switched communications system of Western Union, featuring various bandwidth FDX connections.

buffer—A storage device used to compensate for a difference in rate of data flow when transmitting data from one device to another.

Business Service Instructor (BSI)—A telephone company employee who will teach you how to use your phone system.

busy override—If you're talking on your phone, the operator can interrupt your conversation to announce that another call is waiting (only in business conditions).

B-statement—The section of your phone bill that lists all long distance calls billed to your number.

cable—Assembly of one or more conductors within an enveloping protective sheath, so constructed as to permit the use of conductors separately or in groups.

call blocking—The telephone office attendant may prevent any station from receiving incoming calls. When activated by the attendant, this feature diverts calls to the attendant, or to a recorded message or tone.

carrier—A continuous frequency capable of being modulated, or impresses with a second signal.

carrier system—A means of obtaining a number of channels over a single path by modulating each channel on a different carrier frequency and demodulating it at the receiving point to restore the signals to their original form.

central office—A telephone building housing the equipment needed to keep an exchange operating (an exchange is any 3-digit prefix area). A central office may handle more than one exchange.

Centrex—A sophisticated telephone communications system enabling the customer to receive and make calls without going through a central answering point.

channel, analog—A channel on which the information transmitted can take any value between the limits defined by the channel.

channel, voice-grade—A channel suitable for transmission of speech, digital, or analog data, or facsimile, generally with a frequency range of about 300 Hz to 3400 Hz.

circuit—A means of two-way communications between two points.

class of service—The switching equipment can be programmed to permit the station user to dial calls throughout the world, or the user may be subject to various kinds of restrictions on dialing.

compandor—A combination of a compressor at one point in a communication path for reducing the volume range of signals, followed by an expandor at the other point for restoring the original volume range.

compressor—Electronic device which compresses the volume range of a signal.

cross-bar switch—A switch having a plurality of vertical paths, and electromechanically operated mechanical means of interconnecting any one of the vertical paths with any of the horizontal paths.

cross-bar system—A type of line-switching system using cross-bar switches.

dataphone—Trademarked Bell System (AT&T) data communications equipment and services.

decibel—One-tenth of a bel. Measuring unit of relative strength of a signal parameter, such as power, voltage, etc.

decoded—The reverse of the coding or scrambling process, such as an unscrambled signal.

demodulation—The process of retrieving data from a modulated carrier wave, the reverse of modulation.

descrambled—The decoded or recovered signal resulting from the decoding process.

dial pulse—A current interruption in the DC loop of a calling telephone produced by the breaking and making of the dial pulse contacts of the phone when a digit is dialed.

dial-up—Use of a phone to initiate a call.

digital data—Information represented by a code consisting of a sequence of discrete elements.

digital signal—A discontinuous (discrete) signal.

Direct Distance Dialing (DDD)—A telephone exchange service enabling the user to call others outside his local service area without the assistance of an operator.

disconnect signal—A signal transmitted from the end of a user's phone line to indicate at the other end that the connection should be disconnected.

duplex transmission—Simultaneous, two-way independent transmission in both directions.

Electronic Switching System—Bell System term (abbreviated *ESS*) for computerized telephone exchange.

exchange—Telephone office providing services in a specified area, such as a town, city, village or its environs. Consists of one or more central offices. Serves all numbers within a given three digit prefix area.

exchange trunk—An exchange devoted primarily to interconnecting trunks.

expandor—A transducer which for a given amplitude range or input voltages produces a larger range of output voltages.

facsimile—A system (abbreviated *FAX*) for the transmission of images.

Federal Communications Commission—A board of seven commissioners (abbreviated *FCC*) appointed by the President to regulate all interstate and foreign electrical communication systems originating in the U.S.

Frequency Division Multiplex—A multiplex system (abbreviated *FDM*) in which the available transmission frequency range is divided into narrower bands, each used for a separate channel.

frequency modulation—One of three (abbreviated *FM*) ways of modifying a sine wave signal to make it carry information.

frequency shift keying—Frequency modulation method (abbreviated *FSK*) in which the frequency is made to vary at significant instants.

hertz—A measure of frequency or bandwidth (abbreviated *Hz*), same as cycles per second.

interface—Concept of optimizing the installation of different equipments or systems to ensure their compatibility in terms of connections, signals, levels and impedances.

International Telecommunication Union—The telecommunications agency of the United Nations (abbreviated *ITU*), to provide standardized communication procedures and practices, including frequency allocation and radio regulations on a worldwide basis.

interoffice trunk—A direct trunk between local central offices.

intertoll trunk—A trunk between toll offices in different telephone exchanges.

inversion—Such as in frequency inversion, where the high frequencies are shifted to low, and the low to high frequencies.

masking—The use of tones, noise, music and the like to hide or mask the clear signal.

message unit—Billing by the number of messages made on the telephone line.

modem—A modulator/demodulator to convert a digital output to analog form. A modem also performs the reconversion to digital at the other end.

modulation—Process by which a characteristic of one wave is varied in accordance with another wave or signal. Used in data sets and modems to make equipment compatible with communications facilities.

multiplexing—The division of a transmission facility into two or more channels either by splitting the frequency band transmitted by the channel into narrower bands, each of which is used to constitute a distinct channel (frequency division multiplex), or by allotting this common channel one at a time (time division multiplex).

off-hook—Activating a telephone set.

on-hook—Deactivating a telephone set.

phase modulation—One of three ways of modifying a sine wave signal to make it carry information. The sine wave, or carrier, has its phase changed in accordance with the information to be transmitted.

public switched network—Any switching system that provides circuit switching to many customers. There are four such networks in the U.S.: Telex, TWX, Telephone, and Broadband Exchange.

pulse amplitude modulation—Modulation in which the amplitude of the pulse carrier is varied in accordance with successive samples of the modulating signal (abbreviated *PAM*).

pulse code modulation—Modulation in which the modulating signal is sampled, and the sample quantisized and coded so that each element of intelligence consists of different kinds and/or numbers of pulses and spaces (abbreviated *PCM*).

pulse modulation—Transmission of information by modulation of a pulse or intermittent carrier. Pulse width, position, phase and/or amplitude may be the varying characteristics.

pushbutton dialing—The use of keys or pushbuttons to generate a sequence of digits to establish a telephone circuit connection.

push-to-talk—An operation (abbreviated *PTT*) where the transmitter is keyed by depressing a button on the handset. Also called half-duplex.

recoverable synchronization—The ability to recover or reestablish synchronization automatically when synchronization is lost. Operational feature is scrambling, because disturbances on a communication channel may upset synchronization.

recovered signal—Decoded signal on the descrambled signal.

recovered voice quality—Quality of the voice signal after the process of decoding or scrambling.

regional center—A control center connecting sectional centers of the telephone system together. Every pair of regional centers has a direct circuit group running from one center to the other.

response time—Time a system takes for a given input.

ringdown—A method of signaling subscribers and operators using either a 20-cycle AC signal, a 135-cycle AC signal, or a 1000-cycle signal interrupted twenty times per second.

rolling code—A code that changes with time. Often used to describe a form of bandsplitting that periodically rearranges the frequency displacement of subbands.

scrambled—The encoded or private form of a signal which is unintelligible except when decoded or scrambled.
sectional center—A control center connecting primary centers.
selective calling—Ability of a transmitting station to specify which of several stations on the same line is to receive a message.
simplex—Transmission in one direction only.
spectrum—A continuous range of frequencies, usually wide in extent, within which waves have some common specific characteristic.
subscriber's line—The telephone line connecting the exchange to the subscriber's telephone.
switch hook—Switch on the telephone set supporting the handset.
switching center—A location which terminates multiple circuits, interconnects circuits, and transfers traffic between circuits. May be automatic or semi-automatic.
synchronization—Process of coordinating or bringing into step the transmitted signal and the receiver's decoder used to unscramble or decode it.

tandem office—Office that is used to interconnect the local end offices over tandem trunks in a densely settled exchange area where it is uneconomical for a telephone company to provide direct interconnection between all end offices. Tandem office completes all calls between the end offices, but is not directly connected to subscribers.
tariff—Published rate for service provided by a communications common carrier. Also the vehicle by which the regulating agencies approve or disapprove such services.
Teletype—Trademark of Teletype Corporation's series of different types of teleprinting equipment.
Teletypewriter Exchange Service—An AT&T service (abbreviated *TWX*) in which teletypewriter stations are provided with lines to a central office for access to other such stations.

Telex Service—A dial-up telegraph service.

terminal—Any device capable of sending and/or receiving information over a communication channel.

time division—Coding an analog signal by segmenting it and reorganizing it in the time domain.

time division multiplex—A system in which a channel is established in connecting intermittently, generally at regular intervals, and by means of an automatic distribution, its terminal equipment to a common channel.

toll center—A central office where channels and toll message circuits terminate.

voice frequency/telephone frequency—Audio frequency in the range of 300 Hz to 3400 Hz.

Appendix II
List of Suppliers

incl. Mobile Phone Products & Accessories

A

Action/Dictorgraph Telecommunications: 34 Cambridge St., St. Meriden, CT 06450.
Aerotron Inc.: US One North; Box 6527, Raleigh, NC 27608.
Alaron Inc.: 185 Park St., Troy, MI 48084.
Allied Communications Co.: 18 E. 41st St. New York, NY 10017.
Almotronics: 9815 Roosevelt Blvd., Philadelphia, PA 19114.
Astronautics Corp. of America: 907 S. First St., Milwaukee, WI 53204.
AT &T: 195 Broadway, New York, NY 10007.

B

BBL Industries: 2830 Clearview Pl., Atlanta, GA 30340.
Bohsei Enterprise Co.: 20501 Plummer St., Chatsworth, CA 91311.
Bramco Controls: 1711 Commerce Dr., Box 706, Piqua, OH 45356.
Buckeye Telephone: 1250 Kinnear, Columbus, OH 43229.
Buscom Systems Inc.: 1230 Mt. View Aluiso Rd., Sunnyvale, CA 94086.

C

Capitol Communications: 2817 E. 5th St., Austin, TX 78702.
Car-Phone of Kansas City: 1427 Agnes, Kansas City, MO 64127.
Carterfone Communications: 2639 Walnut Hill Lane, Dallas, TX 75229.
Carter Corp.: 1916-11th St., Rockford, IL 61101.
Chrome Phone Products: 6080 Jericho Turnpike, Commack, NY 11725.
Comex Systems: PO Box 278, Hudson, NH 03051.
Coded Communications Corp: 1620 Linda Vista Dr., San Marcos, CA 92069.
Communications Control Systems: 605 Third Ave., New York, NY 10016.
Communications Products Inc.: 1401-B S. Floyd Rd., Richardson, TX 75080.
Communications Specialists: 11409 Chandler Blvd., N. Hollywood, CA 91601.
Com-Link International Inc.: 101 Lincoln Ave., Pilham, NY 19803.
Communico: 1669 Old Bayshore Rd., Burlingame, CA 94010.
Controlonics Corp.: One Adams St., Littleton Common, MA 01460.
Cordura Marketing Inc.: 10 S. LaSalle St., Chicago, IL 60603.
Coreco Research Corp.: 370 Seventh Ave., Ste. 301, New York, NY 10001.

D

Data-Phone Corp.: 2590 E. Devon Ave., Des Plaines, IL 60018.
Data Signal: 2403 Commerce Land, Albany, GA 31707.
Datotek: 13740 Midway Rd., Dallas, TX 75240.
Davis-Denver Inc.: 2600 S. Parker Rd., Bldg. t, Denver, CO 80232.
Digital Telephone Systems: PO Box 4222, San Rafael, CA 94903.

F

Fan-Tron Corp.: 4023 W. 30th, Little Rock, AR 72204.
Floyd Bell Associates: 1333 Chesapeake Ave, Columbus, OH 43212.

Fold-A-Phone Inc.: 3180 Expressway Dr., South, Central Islip, NY 11722.
Ford Industries: 5001 S.E. Johnson Creek Blvd., Portland, OR 97206.
Fracom Enterprises: 2130 W. Clybourn, Milwaukee, WI 53233.

G
Gimix Inc.: 1337 W. 37th St., Chicago, IL 60609.
G.E. Mobile Radio: P.O. Box 4197, Lynchburg, VA 24502.
Glenayre Electronics: PO Drawer L, Blaine, WA 98230.
Goldberg Co.: 144-19 75th Rd., Flushing, NY 11367.
Graphic Sciences: Commerce Park, CT 06810.
GTE Automatic Electric: 400 N. Wolf Rd., Northlake, Il 60164.
Gutzmer International: 7215 Convoy Crt., San Diego, CA 92111.

I
Integrated Circuits Packaging: 3008 Scott Blvd., Santa Clara, CA 95050.
Integrated Systems Techn.: PO Box 2585, Garland, TX 75041.
International Mobile Telephone Systems; Box 3634, New Haven, CT 06525.
ITT Personal Communications: 159 Terminal Ave., Clark, NY 07066.

J
Jebsee Enterprises: 32-40 69th St., Woodside, NY 11377.

K
Keith-Ian Co., Inc.: 265 Highway 36, W. Long Branch, NJ 07764.
Kustom Data Communications: 1010 W. Chestnut, Chanute, KS 66720.

L
Law Enforcement Associates: 88 Holmes St., Box 128, Belleville, NJ 07109.

Lear Siegler—Electronic Instr. Div.: 714 N. Brookhurst, Anaheim, CA 92803.

Leever Brothers Co.: 525 Warren St., Dayton, OH 54509.

Livermore Data Systems: 2050 Research Dr., Livermore, CA 94550.

LocAtor Tel-Com: Drawer 10609, St. Louis, MO 63129.

M

Magnetic Electronics: 211 Orr Ave., Opelika, AL 36801.

Maxxima Electronics Corp.: 21736 Roscoe Blvd., Canoga Park, CA 91304.

Mieco: 109 Beaver Crt., Cockeysville, MD 21030.

Microelectronic Communications Corp.: 29245 Stephenson Hwy., Madison Heights, WI 48071.

Monroe Electronics: 100 Housel Ave., Lyndonville, NY 14098.

Motorola: 1299 E. Algonquin Rd., Schaumburg, IL 60196.

Mountain West Alarm Supply: 4215 N. 16th St., Phoenix, AZ 85016.

N

Namedroppers: 4170 Admiralty Way, Marina del Rey, CA 90291.

National Telephone & Electronics: 7561 N.W. 77th Terrace, Miami, FL 33166.

Northern Telecom: 640 Massman Dr., Nashville, TN 37210.

North Supply: 10951 Lakeview Ave., Lenexa, KS 55219.

O

Omicrom Data Sytems: 4480 Cote De Liesse Rd., Montreal, Quebec, Can.

Onyx Telephone Center: 12 W. 32nd St., New York, NY 10001.

P

Panasonic: One Panasonic Way, Secaucus, NJ 07094.

Phone Control Systems: 92 Marcus Ave., New Hyde Park, NY 11040.

Phone Devices Corp.: 524 S. Michigan Ave., Chicago, IL 60605.
Phone-Mate: 325 Maple Ave., Torrance, CA 90503.
Phonies Inc.: 19661 N.E. 10th Court, Miami, FL 33179.
Phonetele Inc.: 15414 Cabrito Rd., Van Nuys, CA 91406.

Q
Quasar Microsystems Inc.: 448 Suffolk Ave., Brentwood, NY 11356.

R
Reach Electronics: Box 308; Lexington, NE 68850.
Recoton Corp.: 46-23 Crane St., Long Island City, NY 11101.
Record-A-Call: 16250 Gundry Ave., Paramount, CA 90723.
Regency Communications: 1227 S. Patrick Dr., Satellite Beach, FL 32927.
RF Communications: 1680 University Ave., Rochester, NY 14610.
Remco International: 1723 W. Howard St., Evanston, IN 60202.
Rydax: 76 Belvedere, San Rafael, CA 94902.
Robins Industries Corp.: 75 Austin Blvd., Commack, NY 11725.
Royce Electronics—Phone Div.: 1746 Levee Rd., N. Kansas City, MO 64116.

S
Saxton Products: 215 N. Route 303, Congress, NY 10920.
Seaboard Electronics: 70 Church St., New Rochelle, NY 10805.
Secode Electornics: 908 Dragon St., Dallas, TX 75207.
Smith-Gates Corp.: 1451 New Britain Ave., Framington, CT 06032.
Speedcall Corp.: 2020 National Ave., Hayward, CA 94545.
Standard Communications: PO Box 92151, Los Angeles, CA 90009.
SYS Corp.: 2208 Texas Ave., El Paso,TX 77790.
Sanyo Electric: 1200 W. Artesia Blvd., Compton, CA 90220.

T
Telcer Telephone USA: 319 Lynn Way, Lynn, MA 01901.
Telco Products Corp.: 44 Sea Cliff Ave., Glen Cove, NY 11542,

Teleconcepts: 22 Culbro Dr., West Hartford, CT 06110.
Tele-Tender Systems Corp.: 9 E. 37th St., New York, NY 10016.
Tel Products Inc: 7801 Bass Lake Rd., Minneapolis, MN 55428.
Telephonic Equipment: 17281 Eastman St., Irvine, CA 92660.
Techn. Communications: 56 Winthrop, Concord, MA 01742.
Timekit: 23715 Mercantile Rd., Cleveland, OH 44122.
Technology Applications Corp.: 1101 San Antonio Rd., Mountain View, CA 94040.
Tele-Devices Corp.: N. Country Center, Route 9, Plattsburgh, NY 12901.
Tele-Total: 540 N.W. 165 St. Rd., Miami, FL 33169
Tridar: PO Box 61929, Sunnyvale, CA 94086.
TRS Marketing: 137 E. Savarona Way, Carson, CA 90746.

U

Utility Verification Corp.: 66 Commack Rd., Commack, NY 11725.
Universal Phone Corp.: 1746 Levee Rd., N. Kansas City, MO 64116.

Z

Zoom Telephonics: 65 Franklin St., Boston, MA 02110.

Index

Index

A

Accessories	88
Amplifier	100
Analog channel	44
AT&T	19
AT&T's decorator phone & accessories	114
Automatic telephone dialer	140

B

Bell, Alexander Graham	11
Bell Telephone Laboratories	22
Bel-ringer	97
Brake, percent	35

C

Calls diverter	88
Coherent light	47
Communications, lightwave	46
satellite	51
Computer technology & mobile phone	165
Control, exchange	40

D

Data phone	78
Decorator telephones	110
Dialer, automatic	92
Dials, rotary	35

E

Early facsimile methods	143
Electronic mail	145
phone	75
Exchanges	38
Experiments, Bell	13

F

Fone-a-lert	96

G

Gobbler	100
Government regulations	163
Gray, Elisha	12
GTE automatic Electric Company	22
GTE's Decorator phones & accessories	112

H

Handset	29
Hellos	100
Hooke, Robert	11
Hooking up your phone	42
Hookswitch	31
How your telephone works	27
Huth, G.	11

I

ITT's Decorator phones	118

189

L

Laser	47
Lines, trunk or toll	39
Loop	28

M

Manufacturers & supplier registration	61
Mobile telephones	150
unit to base station calling	154
Modem	45
Modulation, pulse code	45
Multiplexed	44
time division	40

O

Optics, fiber	48
Other decorator phones	120
Owner requirements	63

P

Page, Charles Grafton	12
Picturephone	68
Phone, attache	156
Photodiode, avalanche	50
Photoelectric cell	143
Pusler, dial	33

R

Radio common carrier	164
Recorder	101
Requirements, telephone company	64
Ringer	31
equivalence	62

S

Satellites, active	52
passive	52
Scrambling facsimile communications	146
Silencer, phone	98
Skyphone	162
Soft-touch	88
Speaker phone	72
Station to mobile unit calling	152
Stringent requirements	142
Switching network	37
space division	40
system, digital	40
System, Bell	17
voice analyzer	134
security alarms	137

T

Technique, installation	156
Telegraph, International Telephone	24
Telephone & the computer	56
answering machines	101
cordless	83
service, mobile	148
set	29
Today's telephone	25
Tone dial, pushbutton	37
Touch calling unit	35
Touch-tone	26
Transformer	30
Transmission, digital	45

V

Voice scrambling devices	128

W

Western Electric Company	19
Williams, Charles	15
Wire tap debuggers	132

Y

YAG laser	48
Yellow pages, pocket	99